D0120185

John and Gillian Lythgoe

FISHES OF THE SEA

*The coastal waters of the British Isles, Northern Europe and
The Mediterranean
A photographic guide in colour*

London, Blandford Press

Layout: Frits Stoepman gvn

ISBN 0 7137 0539 6

Printed in Holland by The Ysel Press, Deventer.
Bound in England

Contents

Preface

Until recent years the only live fishes seen by most people were at the end of a fishing line or in an aquarium. Now almost anyone can put on a face mask and see for himself the life beneath the sea. It is unquestionably a strange experience to see for the first time a fish swimming past completely at home in its own environment. Sometimes a fish stops near enough to be scrutinized at close range, sometimes it is only glimpsed in the distance; but in either case it can prove surprisingly difficult to find out what it is called. The whole classification of fishes is based on dead specimens and an accurate identification may require counts of the number of scales and fin rays or the exact proportions of the head These characters are normally displayed on drawings in which the fish is seen in side-view and all the fins are extended. Not surprisingly a living fish seldom obliges with such a display and a swimmer has neither the time nor the knowledge to note the important details of structure, and on his return to the surface, he may only remember that he saw a large shoal of little dark fishes. This sort of information is rarely enough to identify a fish in the normal way but a search through a set of underwater photographs can very often turn up the exact species or at least a near relative. Once the possibilities have been isolated it becomes much easier to use the more formal drawings and descriptions.

In this book we have tried to show photographs taken in life of every common fish (and some quite rare ones) that the coastal fisherman or diver is likely to see. Most of the photographs were taken underwater and show the fishes as they actually appear in nature. However, some very common fishes, notably the Mackerel *(Scomber scombrus)* and the Herring *(Clupea harengus)*, have completely evaded the divers' cameras and instead we have had to rely on aquarium photographs.

During our long search for photographs we have often been asked why we do not concentrate upon aquarium photographs for surely they are easier to obtain and are of better quality. In fact there are several good reasons for preferring the underwater photograph if there is a choice. Coloured fishes may fade very quickly in the aquarium and sometimes they develop unnatural deformities like the over-large lower jaw of the Mackerel. The scenery in a tank is at best the keepers' idea of what it ought to be and it is technically very difficult to photograph a fish through the glass windows of an aquarium. Indeed, in good conditions, an underwater photographer can get results that compare well with those taken on land.

The area covered by this book includes the Mediterranean, North Eastern Atlantic and the North Sea. The salt-water fishes of the Baltic are also included but we have included only those Baltic fishes that are normally thought of as marine, and perhaps we have rather stretched a point in including the Pike *(Esox lucius)* because it is of such great interest to the angler. We have a comparable difficulty at the eastern end of the Mediterranean. Many species have migrated from the Red Sea through the Suez Canal into the Mediterranean and some at least are now established as far east as Cyprus. Some fishes can be identified underwater with certainty, others can be distinguished if a good photograph is available but there remain many that simply cannot be told apart unless the actual body is examined — even then there are arguments and mistakes! Complicated keys to identification have not been given but instead there is an outline drawing of each fish supported by a few words emphasising the most important points. There is also a much more detailed description of each fish which should enable it to be identified with confidence.

The chapter on gobies was written by Dr P. F. Miller who has specialised in this group. These little fishes are notorious for their difficulty and as a consequence our knowledge of them has long been confused and contradictory. The chapter by Dr Miller should go a long way to sorting them out for the non-specialist.

The authors alone are responsible for any omissions and errors that follow. There would undoubtedly have been more had it not been for the generous help and advice from many people. In particular, we would like to thank Dr G. W. Potts, Dr Q. Bone, Mr A. C. Wheeler, Mr J. Hislop, Dr C. C. Hemmings and Mr P. Scoones.

The final word of thanks must go to all the photographers who have contributed to this book. They have allowed us to keep their pictures for long periods whilst the most suitable ones were selected, they have passed on specific requests to other photographers, and they have given expert photographic assistance. Without their skill this book would never have been begun and without their help it would never have been completed.

Introduction

The distribution of marine fishes

All the great land masses of the world are surrounded by a plateau which slopes very gently from the coast to a depth of about 650 ft (200 m.)[1]. The average width of this plateau or 'Continental shelf' is about 30 miles (50 km.) and its average gradient is about 2% (a slope too shallow to be detected by the human eye). These of course are average values and the actual width of the Continental shelf may vary from some 810 miles (1300 km.) off the Arctic coast of N. Siberia to almost nothing where mountain slopes shelve steeply into the sea. In detail the gradient of the bottom is naturally very variable as there are numerous underwater cliffs, reefs and shoals as well as large areas of flat sand and gravel bottom. At the edge of the Continental shelf the bottom slopes much more steeply (between 3% and 6%). This steeply shelving area is the Continental slope which continues down to a depth of some 2,000 m. Below this are true abyssal plains with depths extending to about 6,000 m. About 2% by area of the ocean floor is deeper still and soundings greater than 10,000 m. have been obtained. In general there is a far greater diversity and number of fishes in the waters of the Continental shelf and there is a progressive decline in both as the depth increases.[2] At present diving techniques limit direct observation of fishes to relatively shallow depths and and only those fishes that live in shelf waters are included here.

There are two vast assemblages of marine animals in the world, the one occupies the Indian and Pacific oceans, the other occupies the Atlantic and Mediterranean.[3] There are rather few species common to both areas and these are chiefly migratory open-water species. However, representatives of the same genus are frequently found in both the Atlantic and Indo-Pacific regions. This pattern of distribution is believed to indicate that the Continental barriers between the two regions are of relatively recent date.

The greatest barrier to the distribution of fishes is temperature. The average surface temperature in February varies from 5° C in the North Sea (much lower in the Baltic) and Ireland, to 17° C off the Atlantic coast of Morocco and the Eastern Mediterranean. The corresponding average temperatures for August are about 10° C and 27° C respectively. Despite this wide temperature range it is impossible to point to crisp distributional boundaries in the N. E. Atlantic and Mediterranean, and indeed 35% of the species found north of the Arctic circle are also found in the Mediterranean!

The area covered by this book is much too large to include every species that occasionally enters it from surrounding areas and this has inevitably involved us in some rather arbitrary decisions on what to include. The two areas of greatest difficulty from this point of view are the Baltic and the Eastern Mediterranean. Fig. A

In the Baltic the land-locked ends of the Bay of Bothnia in the north and the Gulf of Finland in the east contain water that is practically fresh with a salt content of about 3 parts per thousand compared to about 35 parts per thousand in the open sea. Some species, such as the Flounder (*Platichthys flesus*) and the Four-horned

1. H.U. Sverdrup, M.W. Johnson and R.H. Fleming, *The Oceans, their physics, chemistry and general biology.* (New Jersey: Prentice Hall, 1942)
2. N.B. Marshall, *Aspects of deep-sea biology.* (London, 1954)
3. S. Ekman, *Tiergeographie des Meeres.* (Leipzig: Akad Verlagsgesellsch., 1935)

Figure A

Sculpin *(Onocottus quadricornis)*, can live in both salt and brackish water and have been included. Fishes that primarily live in fresh water but can tolerate a small amount of salt have generally not been included. One exception is the Pike *(Esox lucius)* which has been included in deference to its great interest to the Baltic angler!

The eastern Mediterranean presents a somewhat different problem and one that would not have occurred before the building of the Suez Canal. The fauna of the Red Sea belongs to the Indo-Pacific and not the Mediterranean-Atlantic. Since the canal was opened however, many Red Sea species have travelled through it into the Mediterranean where they are now established. We have not included the Red Sea Migrants[4] but one of these fishes, the Soldier Fish *Holocentrus ruber* (Forsk) is both colourful and firmly established in Cyprus, where it is often seen hovering almost stationary at the mouth of underwater crevi-

4. A. Ben Tuvia, 'Mediterranean fishes of Israel', *Bull. Sea. Fish Res. Sta.*, (1953), 1-40

ces in moderately deep water. We have also included the Goat Fish, *Pseudupeneus barberinus* since its markings are so distinct. pls 32, 46

Colour and Camouflage

Viewed from the surface the colour of the sea depends very much on the colour of the sky —if the day is clear the sea looks blue, if it is cloudy the sea looks grey. Underwater the colour of the sea depends very little on the weather, and the colour of the water itself is the most important factor.[5] Pure water is blue, and the Mediterranean, which contains very little colouring matter other than the water itself, normally looks blue. On the other hand the waters that surround the coasts of Northern Europe, Britain and Scandanavia contain various colouring agents that stain the water a green or yellow-green colour.

Daylight itself is usually thought of as white and all wavelengths of light are present in about equal amounts. When a narrow band of wavelengths is isolated the eye sees that light as coloured. The short wavelengths give the sensation of violet and blue and the longest give the sensation of red. The colours in order of increasing wavelength (the colours of the spectrum) are violet, blue, green, yellow, orange and red. Pure water absorbs all wavelengths of light but the absorption is much greater at the longer (orange and red) wavelengths. Thus the blue and green wavelengths penetrate deepest into a clear sea such as the Mediterranean. Green northern water is so coloured because the products of vegetable decay which enter the sea *via* the rivers absorb most light at the shorter (blue and violet) wavelengths; the water itself absorbs the orange and red light and the vegetable decay products absorb the violet and blue light leaving the green and yellow light to penetrate deepest into the water. The green colour of the northern seas is also enhanced by the green chlorophyll present in plant plankton which is so abundant at certain seasons.

Coloured objects underwater do not look the same as they do on land because the red light rays are almost non-existent deeper than about 20 m. and in green water the violet and deep blue light is also absent. In practice, deep blue and yellow objects show up best in the Mediterranean and peacock blue and orange objects show up best in the green northern seas.[6] From the practical fish identification point of view the most important of these colour effects is that in the Mediterranean red or pink fishes such as *Anthias anthias* and the Red Mullet *(Mullus surmuletus)* look grey or blue grey instead of the red they appear when brought to the surface.

There are fishes of every hue known to man from the brilliant reds and blues of gurnards to the muddy browns of many bottom living flat fish. Some colours are conferred by tiny pigment-containing cells (chromatophores) in the dermal layer of the skin. These chromatophores are irregular in shape and contain either red, orange, yellow or black pigment. The pigmented area within each chromatophore can expand and contract and this, together with various combinations of chromatophores containing different pigments, is responsible for most of the green, yellow, red, orange, brown and black colours of fishes. The silver, blue and iridescent colours are not produced by pigment but are the result of the interference between different wavelengths of light as they pass through the minute silvery crystals contained in the scales or in special cells called iridocytes.[7] The precise arrangement of these crystals is essential for the correct production of colour and at the point of death the arrangement of crystals begins to break down. This is why the iridescent blue colouring of fishes tends to be lost immediately after death, whilst the flanks of many open water fishes, silver in life,

5. N.G. Jerlov, *Optical Oceanography*. (Amsterdam, London, New York: Elsevier, 1968)
6. J.N. Lythgoe and J.D. Woods, (editors), *Vision in Underwater Science*. (Oxford University Press, 1971)
7. H.R. Norman and P.H. Greenwood, *History of Fishes*. (London: Benn, 1963)

present a rainbow of iridescent colours as they die. Most fishes possess both structural and pigmented colours; the pigment colours being responsible for the dark back and the structural colours for the white and silvery flanks and belly.

Some fishes certainly do have colour vision and that very like our own, and colour is certainly important in the sexual behaviour of some fishes such as the Stickleback *(Gasterosteus aculeatus)*. It probably has other social functions as well, such as the recognition of other individuals of the same species or of those species that live by cleaning parasitic animals and fungi from the bodies of other fishes.

The purpose of camouflage is concealment and many bottom-living forms achieve this by the effective but simple means of burying themselves in the sand. Others, notably the teleost flatfish, are able to change both their pattern and colour to match the bottom. Indeed local fish merchants are able to tell with certainty on what fishing ground a particular specimen was caught.

For really effective camouflage it is equally important to resemble the bottom both in colour and texture and many of the really well camouflaged fishes such as the Scorpion Fishes (Scorpaenidae) have small projections on the skin which resemble the encrusting growth on the rocks where they lie.

Undoubtedly the most sophisticated form of camouflage now known is that shown by the silvery fishes of open water. Underwater most of the light comes from above and very little from below, but the amount of light arriving from all horizontal directions is usually about equal. This being the case a mirror hung vertically in the sea will be almost invisible since the light reflected from it will be equal to the light directly behind it in both colour and brightness. However, a fish is not flat but rounded and a simple mirrored surface would be useless as camouflage. The sophistication lies in the arrangement of the tiny reflecting crystals in each scale which are so arranged on different parts of the body that the whole fish functions as a vertical mirror. This system does not work when looking directly down or directly up at a fish. Perfect camouflage against observation from below is not possible on this system and the fish will always appear as a dark silhouette. However, observation from above is made very difficult by the presence of chromatophores on the back which adjust the brightness (and colour) of the fish to match the dim light coming up from below.[8] The system of mirror camouflage does not work near the surface if there is a clear sky and a low sun for then the sunlight penetrates the water at an angle, and it is said that under these conditions fishes swim with their backs slanted towards the sun.

For perfect camouflage silvery fishes must have the correct position in the sea for once their body tilts they are no longer camouflaged and bright flashes of light are reflected from their flanks. Most of the photographs of silvery fishes in this book were taken with the aid of flash, for by natural light very little would be seen. An exception is the photograph of the Bass *(Dicentrarchus labrax)*. The fish is extremely close to the camera and each scale can be seen but it is evident that the fish is of almost the same brightness as its water background and at a metre or so further off it would be quite invisible.

The art of fish photography

Photographing fish underwater is undeniably difficult since the diver must simultaneously master three different skills. Firstly he must be a practised diver and swimmer so that the fishes are not alarmed by unnecessary noise and the visibility is not ruined by clumsy movements near the bottom; secondly he must be trained in the use of his photographic equipment which is normally encased in bulky waterproof casings, and thirdly he must know the ways of fishes and where to find them.

8. E.J. Denton and J.A.C. Nicol 'A survey of reflectivity in silvery sided teleosts', *J. mar. biol. Assoc. U.K.*, 46 (1966), 685-722

Paradoxically it is this very difficulty that has helped us in compiling this book. On land the natural history photographer tends to search out uncommon species and ignore those that are so common that there is no challenge in photographing them. Underwater it is sufficient challenge to obtain a sharply-focussed, correctly exposed, well framed picture of any fish, and as the commonly seen species are naturally those that present themselves most often it is amongst common fishes that there is the greatest choice of photographs.

Nevertheless some fishes that are commonly caught by the fishermen are rarely seen by the diver and vice versa. The Haddock is an abundant food fish in the water of N.W. Europe yet the civilian diver with his noisy underwater apparatus almost never sees it. The photographed reproduced here **(Pl. 16)** is indeed a rarity and Dr. Hemmings had to lure the fish within camera range by tempting it with broken-up mussels. Other fishes, particularly the smaller rock dwelling species are much easier to photograph than to capture; the Leopard-spotted Goby *(Thorogobius ephippiatus)* was once considered rare in the Mediterranean and was unknown in Northern Europe. Diving scientists and photographers have since shown that it is in fact a reasonably common fish but that its extreme wariness and its habit of taking refuge in cracks and crannies in the rock makes it difficult or impossible to capture by ordinary means.

It will be evident from the preceeding section that two troublesome technical problems that must be met by the underwater photographer are the restricted range of colours and the very dim light. Without the aid of flash, very long exposures and wide apertures may have to be used, making it impossible to obtain a sharp picture. If colour film is used the entire picture will have a blue or green cast which can be very disappointing. The use of a bulb or electronic flash adds extra white light to the scene and many colours such as oranges and reds which are not normally visible underwater are revealed. Really sharp and colourful underwater photographs are invariably taken by flash; yet many photographers prefer the more subtle light and colour that are recorded when only natural light is used.

The description of fishes
FINS (Figs B-E)
These can be divided into two main types, the median unpaired fins and the paired fins.

A. *Median fins*
The primitive form of fin is a continuous fringe which extends along the mid-line of the body from the head around the tail to the vent. In most modern fishes the median fin does not develop equally all round the body and instead there may be one or more dorsal and anal fins and a tail fin. The fins are supported by rays which may be either soft or spiny. Soft rays are branched or jointed whereas true spiny rays are simple in form. Where both types of fin ray are present the soft rays almost always follow the spiny rays. (The Eel Pout, *Zoarces viviparous* is a rare exception).

The median fin rays are sometimes modified into simple spines or have grooves to conduct venom from the poison glands at their base to their tip. Sometimes the modification of the dorsal fin is more extreme or bizarre—the sucker of the Remora is modified from the spiny dorsal rays, as is the long baited filament that the Angler Fish *(Lophias piscatorias)* dangles over its mouth.

The second dorsal fin of the Salmonidae present another type of median fin for here there are no fin rays and the fin is nothing but a flap of tissue and is called the adipose fin.

11

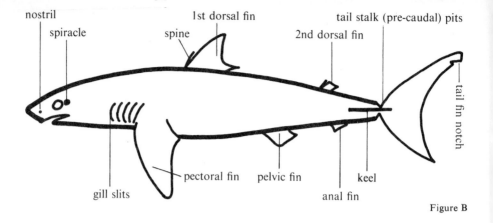

nostril
spiracle
spine
1st dorsal fin
tail stalk (pre-caudal) pits
2nd dorsal fin
tail fin notch
pectoral fin
pelvic fin
keel
gill slits
anal fin

Figure B

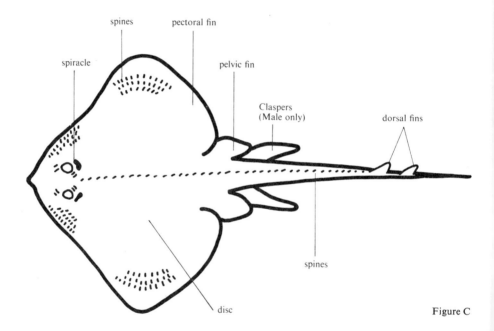

spines
pectoral fin
spiracle
pelvic fin
Claspers
(Male only)
dorsal fins
spines
disc

Figure C

B. *Paired fins*

The pectoral and pelvic fins are the only paired fins. The pelvic fins corresponds with the hind limbs of land vertebrates and the pectorals with the front limbs. The pectoral fins are always attached just behind the head. In the Gurnards (Triglidae) some of the pectoral rays are modified into feelers which carry taste buds on their surface and are used in detecting food.

The position of the pelvic fins is much more variable and they may be completely absent in some fishes such as the Eels (Anguilliformes) and the Pipe Fish (Sygnathidae). In several species the pelvic fins are modified into suckers which enable the fish to cling to rocks. Male sharks and rays have the pelvic fins modified into 'claspers' which are used to hold the female during copulation.

The number of rays in each fin and whether they are soft or spiny is often used in the

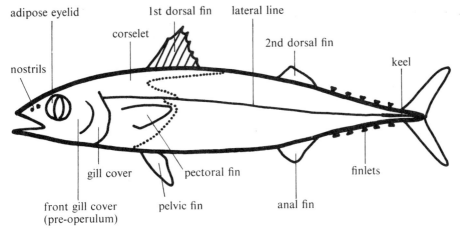

adipose eyelid 1st dorsal fin lateral line corselet 2nd dorsal fin keel nostrils gill cover pectoral fin finlets front gill cover (pre-operulum) pelvic fin anal fin

Figure D

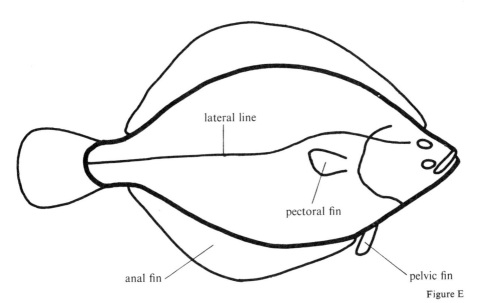

lateral line pectoral fin anal fin pelvic fin

Figure E

description of fishes. The number of fin rays that may be found in any one species is not absolutely fixed but varies within certain limits. In the text that follows these limits are generally indicated and the absence of reported variation probably means that the number of specimens where fin rays have been counted is small.

We have usually only included the counts for dorsal (D) and anal (A) fins. The spiny rays are numbered in roman numerals, the soft rays in arabic. Thus ID X1-X111; 2D 12-16 means the first dorsal fin has between eleven and thirteen spiny rays, and the second dorsal has between 12 and 16 soft rays. A 1V, 8-12 means the anal fin has four spiny rays followed by between eight and twelve soft rays.

SCALES AND SKIN
Most fishes have a covering of scales although in many species the scales are

13

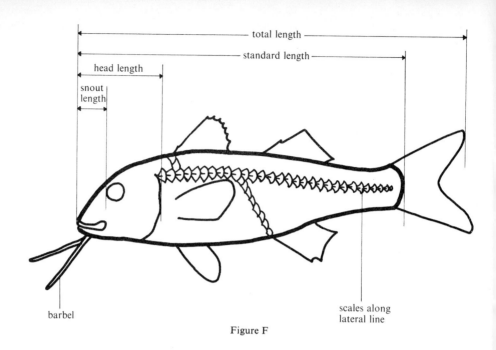

Figure F

much reduced or are absent. In the sharks and rays the scales resemble teeth, having a bony base embedded in the skin and an enamel-covered spine projecting outwards beyond the surface of the skin. Sharks and Dog Fishes are completely covered by denticles. These do not grow with the fish but new ones appear to fill the gaps or to replace those that have disintegrated.

In the Rays the denticles do not cover the whole body but are irregularly distributed over the upper surface especially along the mid-line of the back and on the tail (Fig. C). Denticles are absent in the Electric Rays (Torpedinidae), Sting Rays (Dasyatidae) and Eagle Rays (Myliobatidae).

The bony fishes have scales that are quite different to those of the sharks and rays. The scales are derived from the outer layers of the skin only and grow with the fish. Scales are present in a great variety of shapes and sizes but very broadly speaking the scales are flat overlapping discs with the hind end of each embedded in the skin. Two main scale types can be recognised. Typically in the soft-rayed fishes the hind edge of the scale is smooth (cycloid scale)

whereas in the spiny-rayed fishes the hind edge is finely toothed (ctenoid scale). These distinctions are by no means rigid and many intermediate forms exist, whilst many fishes possess both types of scale.

In some fishes such as the Herrings (Clupeidae) the scales are only superficially embedded and become detached with only the slightest handling, whereas in others such as the flat fish and the Eels the scales are deeply embedded in the skin and do not overlap.

Since the scales of bony fishes grow with the fish and new ones are only produced to replace lost ones, the number of scale rows is a useful character to distinguish fishes. Moreover since fishes grow faster at some seasons of the year than at others, the pattern of 'growth rings' on the scales can give an indication of age. A similar estimate of age can be obtained from the rings in the stones of the inner ear (otoliths). These otoliths show some variation in shape from species to species but their use in identification is outside the scope of this book.

The number of scales along the lateral line of bony (teleost) fishes and the number of rows of scales above and below the lateral line (Fig. F) can be important aids to identification. In the Wrasse (Labridae) the number of rows of scales on the cheek below the eye is an important character.

TEETH

The teeth are adapted in shape to serve the nutritional requirements of the fish. Thus long sharp canine-like teeth are used for catching and holding live prey, chisel-like teeth are used for browsing algae from rocks and strong flattened teeth are used for grinding up shellfish.

The Hagfish and Lampreys do not have biting jaws but their sucker-like mouth which they use to attach themselves to other fishes is studded with horny 'teeth'. The tongue is also armed with these horny 'teeth' and is used to rasp the flesh from their prey.

True teeth are found in cartilagenous and bony fishes. The teeth in cartilagenous fishes are only found in the jaws but in the bony fishes they may be also situated on the tongue, on the roof of the mouth (palatine and vomerine teeth) and on the inner margin of the gill arches (the pharyngeal teeth).

In the Hagfish and Lampreys the teeth are replaced by new ones developing beneath the ones presently in use. In the sharks and rays the oldest and most developed teeth are in the outside of the jaw and the underdeveloped young teeth are ranged in the inner files. As an old tooth is lost, a younger tooth moves up to take its place. This regular replacement of teeth is not found in the bony fish where replacement teeth grow up at the base of existing ones or in gaps between them.

THE LATERAL LINE (Figs. D, E, F.)

Fishes are very sensitive to the vibrations and minute pressure changes in the water which are set up by their own motions and by other fishes and objects in the water. In most fishes the sense organs that detect these pressure changes (the neuromast organs) are set in tubes running beneath the surface of the skin and these connect to the outside water through pores in the skin. The course of these neuromasts along the flanks of the fish is marked by the lateral line. On the head the line branches into three, one passing above the eye to the snout, the second below the eye to the snout and the third below the eye to the jaw. The lines on the head are usually less obvious than the lateral line itself and are often sub-divided. The tubes are often situated deeper in the head and their course can only be detected by the minute pores that communicate with the surface.

THE EYES

The eyes of fishes have a basic structure very similar to those of mammals, and some at least have colour-vision[9]. One difference is that the cornea, being in direct contact with the water, is unable to help focus the light on the light-sensitive back of the eye and instead all the focussing power of the eye lies in the lens which is thus almost spherical. Another difference is that fishes have no true eyelids. However in many fishes, especially in free-swimming fishes such as Mackerels and Herrings, there is a flap of transparent tissue (the adipose eyelid, fig. D) that covers the eye except for a vertical slit across the centre.[10,11] The light underwater is polarized due to the light scattering action of the water and the adipose eyelid, which acts as a polarizing filter, may help the fish to see through the water. In some sharks the eye is protected by a membrane (the nictating membrane) which can be pulled down across the eye like a blind.

The size of the eye in relation to other parts of the body may be important in naming a fish. The size of the eye is usually

9. K. Tansley, *Vision in vertebrates*. (London: Chapman and Hall, 1965)
10. G.L. Walls, *The vertebrate eye*. (New York and London: Hafner, 1963)
11. K.W. Stewart, 'Observations on the morphology and optical properties of the adipose eyelids of fishes', *J. fish. res. Bd. Can.*, 19 (1962), 1161-1162

taken as the distance between the two opposite bony edges of the eye cup (orbit diameter).

NOSTRILS (Figs. B, D.)

In the fishes the sense of smell is usually well developed, especially in the sharks and rays. The organs that are responsible (the olfactory organs) are located in the head and their opening to the outside world is through the nostrils. In the sharks and rays the nostrils and olfactory organs are on the under side of the head and the nostril can be closed by special flaps. In the bony fishes there are normally two pairs of nostrils and since their position on the head varies from species to species they can be helpful in the identification of fishes.

Both the taste buds and the olfactory organs detect the presence of chemical substances in the water, and as such are very closely allied. However, the two senses are anatomically distinct and the nerve routes carrying information from the receptors to the brain are different in the two senses.

TENTACLES

Some fishes have little outgrowths from the skin which may be simple filaments or quite elaborate branched structures. They are often situated above the eyes or on the rim of the nostrils.

BARBELS (Fig. F)

The taste buds in fishes are not always confined to the mouth but may be found on various parts of the body. In particular many fishes, especially bottom feeders, have concentrations of taste buds on the barbels which are concentrated around the mouth or on the chin. It is a common sight to see fishes such as the Red Mullet *(Mullus surmuletus)* feeling over the bottom with their barbels.

The classification of fishes

The whole art of biological classification is to arrange all living things into groups of related forms[12], The smallest group is the species, and normally an individual animal or plant can only breed with another of the same species; thus Man is one species, the Gorilla is another. All species that are judged to be very closely related are grouped into one genus. For instance, amongst the Sea Breams the two species *Diplodus vulgaris* and *Diplodus annularis* are considered to be sufficiently closely related to be included together into one genus. Whereas another Bream such as *Charax puntazzo* is placed in another genus chiefly because of its different teeth. However, authorities agree that all the Sea Bream are in fact fairly closely related and they are all grouped into one family, the Sparidae. Classification into families and genera is largely a matter of convenience and if there are a large number of closely similar fishes such as the Gobies, family and generic divisions tend to be made on much smaller differences than in groups where there are far fewer forms.

The familes are in turn grouped together into orders, orders are grouped into classes, classes into phyla and finally the phyla are placed in either the animal or plant kingdoms. Nevertheless some very simple forms of life can not be placed with confidence into either kingdom.

Fishes are placed in the same phylum (Chordata) as such apparently diverse forms as Sea Squirts (Tunicates), Amphibia, Reptiles, Birds and Mammals. Other large and important phyla are the Insects, Molluscs and the Echinoderms.

A synopsis of the phylum Chordata is set out below.

Phylum Chordata.

Animals which at some stage of their life history possess a nerve cord running along the dorsal part of their body. The phylum Chordata is divided into the following sub-phyla:

12. A.S. Romer, *Vertebrate Palaeontology.* (University of Chicago Press, 1945)

Sub-Phylum
1. Hemichordata (e.g. *Balanoglossus*)
2. Cephalochordata (e.g. *Amphioxus*)
3. Urochordata (e.g. Sea Squirts)
4. Vertebrata

The members of the sub-phylum Vertebrata (or Craniata) all have vertebrae developed to some degree and an elaboration of the nerve cord at one end housed in a head or cranium. The Vertebrata are divided into two superclasses:

Superclass 1. Agnatha
Class Cyclostomata
    Vertebrates with no proper jaws
    (e.g. Hagfish, Lampreys)
Superclass 2. Gnathostomata
    Vertebrates with jaws. They may be divided into the following classes:
Class 1. Elasmobranchii
    (= Chondrichthyes). Fishes with a basically cartilagenous skeleton (e.g. skates, rays, sharks, electric rays, guitar fish).
Class 2. Osteichthyes.
    Fishes with a basically bony skeleton (e.g. Sturgeons, Lung Fish, Bowfins and modern (teleost) fishes).
Class 3. Amphibia
Class 4. Reptiles
Class 5. Aves
Class 6. Mammalia

The naming of fishes
In the days when everyone remained in their home district there were few problems over the naming of fishes—those that were interesting or important enough were given a name and the rest were ignored. But difficulties arose when people began to travel further afield, for perhaps a few kilometres along the coast a completely different set of names are given to the same fishes so that even within one country all the names are different. To make matters worse colonists to other lands tended to name the new, unfamiliar, fishes by the names of the fishes in the old country. The answer to this chaotic situation was given by Linnaeus in the 18th century. He proposed that each species of animal and plant should have two names the first (generic) name belonged to the group of closely related organisms, the second (specific) name belonged to that particular species alone. This system was proposed at a time when Latin was an international language and the technical system of names has remained international ever since.

There are strict rules about the use of technical names to standardise their use as much as possible. The name first given to the fish is the one that is in principle adopted and sometimes a name in common use is changed because a species has already been satisfactorily described under another name. The name of each fish is followed by the author who conferred it; but it often happens that a later authority decides that the generic name does not reflect the true relationship of the fish and he changes it to something more appropriate. In this case the authority who first described the fish under its original name appears in brackets. The result of all this is that the same fish may have different names in different books according to what seemed correct at the time. There is no answer to this except to list the principle names given in the most commonly used books.

Literature
There are an enormous number of specialist books and articles on fishes and this is not the place to attempt a comprehensive list. Instead we have indicated below some of the more important works and anyone wishing to learn more about some aspect of the subject will be able to trace the specialist literature from them. Most of the original research on fishes is published in the form of individual papers which appear in a large and growing number of scientific journals. The task of keeping relatively up-to-date is made possible by the abstract journals e.g. Biological Abstracts, The Zoological Record (Pisces) and Aquatic Biology Abstracts. Those wishing to know more about Mediterranean fishes are in a particularly fortunate position since Professor Bini's book is now published, and it is both detailed and copiously illustrated.

MEDITERRANEAN FISHES

G. Bini, *Atlante dei Pesce delle Coste Italiane*. (8 vols. Rome: Mondo Sommerso, 1967)

W. Goëau-Brissonniere, *Atlas des Poissons des côtes Algériennes*. (Imbert E. Alger, 1956)

R. Reidl, *Fauna und Flora der Adria*. (Hamburg, Berlin: Paul Parey, 1963)

NORTHERN EUROPE

E.A. Andersson, *Fiskar och Fiske i Norden*. (2 vols. Stockholm, 1954)

F. Day, *The Fishes of Great Britain and Ireland*. (2 vols. London: Williams and Norgate, 1880-1884)

M. Poll, *Faune de Belgique-Poissons Marins*. (Bruxelles, 1947)

A.C. Wheeler, *The Fishes of the British Isles and North West Europe*. (London, Melbourne, Toronto: Macmillan, 1969)

Class: CYCLOSTOMATA

The **Lampreys** and **Hagfishes** are very primitive vertebrates which superficially resemble eels. They have no bones in the true sense but instead are strengthened with separated nodules of cartilage. There are no scales and the body is very slimy. Instead of jaws there is a round or slit-shaped mouth armed with horny teeth. The single nostril is on top of the head and it opens into from 1 to 15 pairs of simple gill openings. There are two orders the *Myxinoidea* (Hagfish) and the *Hyperotreta* (Lampreys).

Order: MYXINOIDEA

Family:

MYXINIDAE

Purely marine. The Hagfish have slit-like mouths with barbels and poorly developed fins. The eyes are degenerate.

Hagfish
Myxine glutinosa L.

Barbels around snout;
1 continuous fin around body.
35 cm.

North East Atlantic; north North Sea; Mediterranean; Atlantic coast of North America.

Body eel-like; *eyes* are not visible and very degenerate: *nostril*, 1 at extreme tip of the head; *barbels*, (see diagram) one pair each side of the nostril and another pair, about twice as large as the others, each side of the mouth; *gill slits* are present as one pair of small round openings on abdomen; *skin* smooth, no scales; *fins*, 1 fin which is narrow and starts about 3/4 of the way along the body, runs around the tail and ends about 1/2 way along the underside.

Yellowish, greyish or reddish. darker on the back, lighter underneath. Usually 30-35 cm. sometimes up to 45 cm.

Hagfish are extremely sensitive to conditions and are only found in areas of high-medium salinity (31-34 parts per thousand), low temperature 10-13°C and low light intensity. The eyes are very poorly developed but the skin is somewhat sensitive to light. They are found on or very near muddy bottoms where they bury themselves into the mud. Their burrows are marked by small mounds with a hole at the top for the fish to enter. Rarely found above 25 m. usually between 30-500 m. They feed on any dead or dying animals by fastening themselves to the skin, often to the gills or on any injuries that the fish may have. They then devour the whole of the animal except for the skin and bones. When they find some particularly tough meat they are able to tear off the mouthful by tying their body into a knot and pulling the mouthful through the knot. Each fish has both sets of reproductive organs but only one set in any fish reaches maturity. The eggs, which are horny with many filaments at each end, are laid throughout the year. There is a row of mucus sacs along each side of the body that produce a prolific quantity of slimy mucus. A single fish is capable of filling a 2 gallon bucket with slime and water in only a few seconds.

Order: HYPEROTRETA

Family:

PETROMYZONIDAE

All **Lampreys** breed in fresh water but some species spend part of their adult life in the sea and are thus included here. The mouth is circular or funnel-shaped and there are no barbels. Eyes are present in the adult but are rudimentary in the larvae.

Sea Lamprey

Pl. 138

Petromyzon marinus L.

Adults with dark mottling:
mouth disc (see diagram)
90 cm.

North west European coasts, Iceland, Atlantic coast of North America, occasionally Baltic, Mediterranean.

Body eel-like, front half round and then gradually becoming flattened sideways towards the tail; *eyes* small; *mouth* a sucker with a fringed edge, oval when in use and slit-like when closed; *nostril*, 1 at the level of the eye on the back; *teeth* (See diagram), the arrangement of the teeth in concentric rows with one large double-pointed central tooth is an important feature for precise identification; *gill slits* 7 pairs; *scales* none, skin smooth; *fins* 2 dorsal fins separate but sometimes nearly touching. The second fin is larger than the first and has a distinct notch between it and the tail fin.
Back olive-brown to yellow-grey with a black mottling. Underside greyish. Young fish may not have the black markings.
Up to 90 cm.

Found at all depths to 500 m. and frequently near the mouths of rivers. The majority of their life is spent in fresh water. In the breeding season the male fish develop a ridge along their backs and the females a fold on their lower surface in front of the tail fin. The adults move into the rivers during the spring and make their nests on the gravel bottoms of fast flowing streams. Activity is restricted to the night and the day is spent attached to a rock each by their sucker. Spawning occurs during May and June, and up to 200,000 eggs may be laid. Both the male and female fish attach themselves to a rock by their suckers and the male twists around the female. After spawning they die. The young fish are very different from the adults and live in mud for about 6 years after which they change into the adult form and migrate to the sea.

Lampreys feed on living fish by attaching themselves to the skin of the prey, rasping through it and sucking out the blood. In some areas (the Great Lakes in Canada) they are a serious menace to fisheries. In the past they were considered a great delicacy and Henry I is said to have died from "his having made too plentiful a repast on these fishes".

Lampern
Lampetra fluviatilis (L.)

Uniform colour:
mouth disc (see diagram)
50 cm.

Mediterranean, coasts and rivers of Britain and Europe north to Norway.

Body eel-like; *eyes* present but small; *mouth* an oval shaped sucker; *teeth* characteristically arranged (see diagram) and are important for precise identification; *gill slits* 7 pairs; *scales* none, skin smooth; *fins*, 2 dorsal fins separate but may be joined together during the breeding season. The anal fin is enlarged during this time especially in the male.

A uniform pale brown, bronze brown, grey or green on the back shading to white underneath. The colour is very variable with locality and season. Iris of eye blue.
Up to 50 cm.

Found off-shore particularly in brackish water and river estuaries. During the autumn the fish migrate into rivers and spawning occurs at a temperature of 11°C. The male excavates a nest amongst the sand of the river bed in fast flowing water. The female attaches herself to a stone by her sucker and the male attaches himself to the female above the eye and winds himself around her. Spawning usually occurs in groups of 10-50 individuals. Up to 40,000 eggs may be laid after which the parents die. The young fish which are completely different in appearance to the adults live in the mud and after about 5 years they change into the adult form and migrate to the sea. They spend at least 1 year in the sea and never move far from the coast where they feed parasitically on fish. When they enter the rivers to breed feeding stops.

At one time they were considered a delicacy and are excellent stewed.

Class: **CHONDRICHTHYES** (Cartilagenous fishes)

The **Sharks**, **Skates**, **Rays** and **Chimaeras** (rabbit fishes). Included in this class are all fishes that have a basically cartilagenous skeleton, no true fin rays and a mouth underneath the head. There is a pair of nostrils and a series of exposed gill openings. The skin is covered with hard dermal denticles or placoid scales. These are tooth-like structures with a bone-like base embedded in the skin and an enamel covered spine which is directed backwards and is exposed. It is these that give the characteristic rough surface to these fishes.

Fertilization is always internal and the mature males have special copulatory organs (claspers) modified from the pelvic fins. Each clasper has a cartilagenous skeleton and a canal running along its length. During copulation the female is grasped by the claspers and the seminal fluid is introduced into her cloaca through the canals. The young may pass their early life in one of three ways. In the oviperous type the eggs, which are usually encased in a horny capsule, are laid directly into the sea soon after fertilization. In the ovoviviperous type the eggs are not laid into the sea but are retained in the oviducts. The eggs break just before birth and the young are born alive into the sea. In the viviperous type the young are nourished through a 'placenta' in the mother, and are also born alive.

The cartilagenous fishes in this book fall within two orders, the Pleurotremata (Sharks and Monkfishes) and the Hypotremata (Rays).

Order: PLEUROTREMATA

The **Sharks** and **Monkfishes**. Front margin of each pectoral fin is free. Gill slits on the side of the head.

The Sharks are the most dangerous group of fishes. They are a serious menace to swimmers in some places, notably in Australia. However they do not seem to attack swimmers and divers in waters further north than the Mediterranean and even there it is extremely rare for a shark to attack.

They must always be treated with great respect and the diver who gets blasé about them and tries to provoke them, or who swims too close, is asking for trouble.

Family:

HEXANCHIDAE

Medium to large sharks with 6-7 gill slits, one dorsal fin without spines, a well marked upper lobe to the tail fin and an anal fin. The eyes have no nictating membrane and the spiracle is rounded. They are viviparous.

Six-gilled Shark *Hexanchus griseus*   p. 23   **Seven-gilled Shark** *Heptranchias perlo* p. 23

1 Dorsal fin;
6 gill slits.
5 m.

1 Dorsal fin;
7 large gill slits.
3 m.

Six-gilled Shark; Grey Shark p. 22
Hexanchus griseus (Bonnaterre)

Mediterranean, Atlantic, English Channel, North Sea and throughout all temperate seas.

Body slender; elliptical; *head* wider than body; *snout* rounded; *eye* large and oval with no nictating membranes; *teeth* (see diagram) different in each jaw, those in the upper jaw are triangular and those in the lower jaw are pointed with cusps at each side; *gill slits* 6, gradually decreasing in size from the head backwards and they are all situated in front of the pectoral fin; *fins*, 1 dorsal fin situated between the pelvic and anal fins. Tail fin large and notched near the end of the upper lobe.

The colour can vary from light to dark brown, grey or reddish. The underside is lighter and the lateral line is usually distinct as a lighter stripe.

Up to 5 m. The females are usually longer than the males.

Solitary deep water sharks found down to 2000 m but usually between 200 and 1000 m. Occasionally they may move into more shallow water particularly in their northern ranges. They feed principally on fish but also on crustacea.

The young are born alive between 40-60 cm. in length and in very large numbers. Up to 100 embryos have been reported and 47 confirmed. Sexual maturity is reached at about 2.1 m.

Seven-gilled Shark p. 22
Heptranchias perlo (Bonnaterre)

Mediterranean, Central Atlantic from Portugal to Cuba.

Body long and rather slender; *head* narrow; *snout* pointed; *eye* large (wider than the distance between the nostrils) and oval; no nictating membranes; *teeth* (see diagram) different in each jaw. Those in the upper jaw have one point and those in the bottom jaw have many; *gill slits* 7, they are very large and continue under the throat. They decrease in size from the head backwards and are all in front of the pectoral fin; *fins* 1 dorsal fin between pelvic and anal fins. Tail fin large and notched near the end of the long upper lobe.

Greyish back, lighter underside. The front edges of the pectoral fins are edged with white and the tip of the dorsal fin is black. There are two white spots on the dorsal fin. The upper edge of the tail fin is black with the extreme rear edge and lower edge bordered in white.

Up to 3 m.

Found in deep water down to 300 m. They are fish eaters and probably live near the bottom. The young are born between 25-28 cm. in length and up to 20 in number.

23

Family:

ODONTASPIDIDAE

Medium to large **sharks** with 5 gill slits, 2 dorsal fins and an anal fin. The upper lobe of the tail fin is large. The head is flattened but the snout is conical. There is a small spiracle. Each scale has 3 keels. No nictating membrane. Viviparous.

Sand Shark
Odontaspis taurus (Rafinesque)

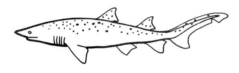

Grey brown with brown to yellow irregular spots.
3 m.

Mediterranean; East Atlantic from Morocco to South Africa and all warm seas.

Body long and fairly stout; *head* small, flattened from above; *snout* pointed; *eyes* small and round; *teeth* (see diagram) similar in each jaw, pointed with cusps on each side; *gill slits* 5, all situated in front of the pectoral fins; *scales* small and scattered over the surface; *fins*, 2 dorsal fins nearly equal in size, the 1st completely in front of the pelvic fins. Tail fin large with a deep notch on the underside of the upper lobe. Pectoral fin longer than in the Fierce Shark (*Odontaspis ferox*).
Greyish or brownish on the back becoming lighter underneath.

Brownish-yellow irregular spots are scattered over the back and sides. The spots are more clearly marked in young fish than in adults.
Up to 3 m.

Deep water sharks which sometimes come into the more shallow water near coasts. They normally live over sand but spend their life continually swimming. They are able to swallow air and keep it in their stomachs which they then use as an air bladder to help them retain buoyancy in the water.
Predominantly a fish and cephalopod-eater they will attack any animal that attracts them. Viviparous, the young are born in winter.

Fierce Shark
Odontaspis ferox (Risso)

Dark rear edge to fins.
2 m

Mediterranean, East Atlantic to Madeira and the Gulf of Gascony.

Body long and fairly stout; *snout* prominent and more or less pointed; *eyes* round and small withouth nictating membranes; *teeth* are similar in both jaws, and are pointed with cusps on either side; *gill slits*,

24

5, all situated in front of the pectoral fins; *fins* 2 dorsal fins, the front edge of the 2nd fin situated in front of the anal fin. Tail fin long (total length of fish 3-3 1/2 times tail), upper lobe elongate and notched near the end.

Dark-grey to black on the back shading to light grey underneath. The 1st dorsal, 2nd dorsal, anal and pelvic fins have black rear margins. The lower edge of the tail fin is also black.

Up to 2 m., occasionally 4 m.

Deep water sharks which approach shore during the summer. Very little is known about their activities but stomach contents show that they are carnivorous.

Family:

ISURIDAE

Medium to large oceanic **sharks.** There are 5 gill slits. 2 dorsal fins and an anal fin which is about the same size as the second dorsal fin. The lower lobe of the tail fin nearly equals the upper giving a 'moon shape'. The snout is pointed and conical. The tail stalk is strongly flattened from above and below. There is a lateral keel on each side which extends onto the fin.

Mako *Isurus oxyrinchus*

Pointed head;
slender, moon-shaped tail;
1st dorsal fin behind rear edge of pectoral fin.
4 m.

White Shark *Carcharodon carcharias*   p. 26

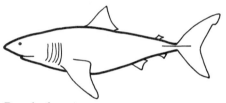

Deep body;
moon-shaped tail;
triangular teeth with saw edges.
6 m.

Porbeagle *Lamna nasus*   p. 26

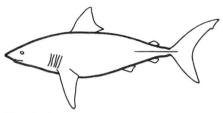

Deep body;
moon-shaped tail;
1st dorsal fin immediately behind pectoral fin.
3 m.

Mako   pl. 139
Isurus oxyrinchus Rafinesque

Mediterranean; throughout tropical and temperate Atlantic north to Scotland; western English Channel.

Body elongate and more slender than Porbeagle (*Lamna nasus*); tail stalk with keels present on either side and well developed upper and lower pits; *head* slender;

25

snout pointed; *teeth* (see diagram) similar in each jaw, long and pointed with no cusps; *gill slits*, 5, long and all situated in front of the pectoral fins; *fins*, 2 dorsal fins the first situated behind the pectoral fins and the second, which is much smaller, originates just in front of the anal fin. Tail fin moon-shaped but with the upper lobe longer and notched.

Greyish or bluish-grey above, whitish underneath.

Up to 4 m.

Solitary and only found in shallow open water, rarely below 18 m. Often confused with the Porbeagle (*Lamna nasus*). They have often been observed jumping out of the water and are a much faster fish than the Porbeagle. They are carnivorous and feed on shoals of small fishes. Sexual maturity is reached at a length of 2 m. in the male, rather longer in the female.

Porbeagle; Mackerel Shark p. 25
Lamna nasus (Bonnaterre)

Mediterranean; throughout Central Atlantic; East Atlantic north to Norway and Iceland, south to Morocco; North Sea; west Baltic; English Channel.

Body deep and robust; *eye* large; *keels* are present each side of the tail stalk and smaller ones on the tail fin; *pits*, there are distinct upper and lower pits immediately in front of the tail fin; *snout* rounded; *teeth*

(see diagram) pointed with cusps on either side and similar in both jaws; *gill slits*, 5, long but they do not reach the under surface. They are all situated in front of the pectoral fins; *fins*, 2 dorsal fins; the 1st immediately behind the base of the pectoral fin and the 2nd, which is much smaller, situated above the anal fin. Tail fin moon-shaped but with the upper lobe notched and longer.

Dark grey to grey-blue on the back shading to white underneath.

Up to 3 m., rarely 4 m.

Found from shallow water down to 150 m. Often confused with the Mako (*Isurus oxyrhinchus*). The Porbeagle is often encountered in colder water than other sharks. They are strong swimmers and feed principally on shoals of fish especially mackerel, which small groups of sharks sometimes follow into shore. The young are born at a length of 50-60 cm., probably in summer and only in small numbers. Considered excellent sporting fish and sometimes found on sale as 'swordfish'! However, they are an excellent food fish in their own right.

White Shark pl. 140 p. 25
Carcharodon carcharias (L.)
= *Carcharodon rondeletii* Müller & Henle

Mediterranean; Atlantic north to Biscay; all temperate and tropical seas.

Body deep and elongate with keels each

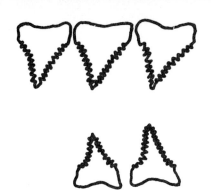

fins; *fins* 2 dorsal fins, 1st triangular and behind the pectoral fin; 2nd smaller and immediately in front of the anal fin. Tail fin large and moon-shaped with lobes nearly the same size; upper lobe notched. Pectoral fins large and long.

Grey-brown to slate blue on the back shading to light grey underneath. The pectoral fins have a dark spot at their tip on the under surface.

Up to 10 m., more usually between 3-6 m.

side of the tail stalk; *snout* rounded; *teeth* (see diagram) triangular with saw edges and are characteristic of this shark; *gill slits*, 5, long and all situated in front of the pectoral

Found in the open sea either singly or in small groups. These are very voracious sharks and known to attack bathers. Feed on fish, turtles, seals, refuse from ships etc.

Family:

CETORHINIDAE

There are several species of **Basking Shark** in temperate seas but only one in our area. They are distinguished by having horny gill rakers. Otherwise they resemble the Isuridae except that they have very numerous minute teeth. Viviparous.

Basking Shark
Cetorhinus maximus (Gunnerus)
= *Selache maxima* Cuvier

Pls. 141, 142

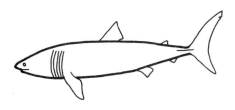

5 very large gill slits;
minute teeth;
young specimens have an elongate snout.
12 m.

Mediterranean; East Atlantic north to Iceland and Norway south to Madeira.

Body deep and elongate with keels present on each side of the tail stalk; *snout* blunt. In young fish the snout is greatly enlarged and moveable; *eyes* small; *teeth* very small; *gill slits*, 5, very long extending from the back to the under surface and are characteristic of this shark; *fins*, 2 dorsal fins, 1st situated between the pectoral and pelvic fins, 2nd immediately in front of the anal fin. Upper lobe of the tail fin larger and notched near rear end.

Brown or grey on the back shading to light grey underneath.

Up to 15 m., usually between 3-12 m. Can weigh over 3 tons.

Basking Sharks are primarily inhabitants of open water but during the summer they are thought to make a migration towards shore where they are occasionally seen by divers. They hibernate in the winter. A gregarious species usually seen in groups of between 12-15 individuals but sometimes up to 60. When seen they are usually 'basking' or swimming slowly near the surface feeding with their mouths open and their snout and dorsal fin and tail tip just emerging above water. This habit has almost certainly given rise to 'sea-serpent' stories in the past as have their decaying carcasses washed up on beaches. Basking sharks are plankton eaters, the gill system filters the plankton from the sea water and up to 1000 tons of water may be filtered each hour. Sexual maturity is reached when they have attained 4-6 m.

Family:

ALOPIIDAE

The very long upper lobe to the tail fin of the **Thresher Shark** is a family characteristic. There are two dorsal fins, the second being the smaller and about the same size as the anal fin. The tail stalk is compressed laterally. There are spiracles and the eyes have nictating membranes. Viviparous.

Thresher Shark
Alopias vulpinus (Bonnaterre)

Very long upper tail lobe.
3 m.

Mediterranean; Atlantic; all temperate and tropical seas.

Body slender with no keels on the tail stalk. A pit is present on the top of the tail stalk immediately in front of the tail fin; *snout* blunt; *eyes* large; *teeth* small triangular and smooth (see diagram); *gill slits* 5, very small; *fins* tail fin with upper lobe greatly elongated and measures nearly half the total length of the fish; 2 dorsal fins the 1st situated between the pectoral and pelvic fins. Pectoral fins long.
Dark blue, grey, brown or black on the back shading to white underneath. The underside of the snout and the pectorals are dark.

Up to 3 m.
Adult fish are found in the open sea but younger specimens are encountered near shore. Usually solitary but sometimes found in pairs and they have been reported as hunting in pairs. The two sharks will encircle a shoal of small fish and thrash the water with their tails forcing the fish into a smaller and smaller area where they can be attacked more easily. Occasionally fish and sea birds are stunned by the force of the tail. The young are born in summer in small numbers.

Family:

SCYLIORHINIDAE

These are the well known **Dogfish** which are in reality small sharks. They are very widely distributed but always live near, or on, the bottom. Usually two dorsal fins set near the end of the body. The upper lobe of the tail fin is nearly horizontal. There are 5 gill slits and spiracles are present. The teeth are small and numerous with a variable number of points. The nostrils are connected to the lip edge by a groove and this is a useful character for identification. Oviperous.

Lesser Spotted Dogfish; Rough Hound
Scyliorhinus canicula

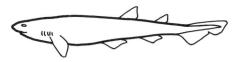

Lower lobe of tail fin larger than upper and not directed upwards;
2nd dorsal fin immediately behind anal fin;
type of nasal grooves (see text).
75 cm.

Large Spotted Dogfish; Nurse Hound   p. 30
Scyliorhinus stellaris

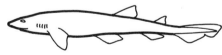

Lower lobe of tail fin larger than upper and not directed upwards;
2nd dorsal begins above anal fin;
type of nasal grooves (see text).
1 m.

Black-mouthed Dogfish   p.30
Galeus melastomus

Lower lobe of tail fin larger than upper and not directed upwards;
blotches along back, arranged in rows.
90 cm.

Lesser Spotted Dogfish; Rough Hound
pl. 1
Scyliorhinus canicula (L.)
= *Scyllium canicula* Cuvier

Mediterranean; East Atlantic north to Norway south to Senegal; North Sea; English Channel.

Body slender and elongate gradually tapering towards the tail; *snout* rounded; *eyes* oval, without nictating membranes but with a thick fold of skin on lower margin; *nostrils* present and connected to the mouth by grooves which form a flap. The simple shape of this flap makes the Rough Hound easily distinguished from the Nurse Hound (*Scyliorhinus stellaris*) (see diagram); *gill slits*, 5, small, the last two situated over the pectoral fin; *skin* rough; *fins* 2 dorsal fins, the 2nd dorsal situated immediately behind the anal fin, the 1st behind the pelvic fin. Tail fin only slightly directed upwards, the lower lobe larger than the upper.
Back brown or reddish-brown, grey or yellow-grey with many small brown and black spots, sometimes with small white spots. Sides lighter and gradually shading to grey-white underneath.
Up to 75 cm.

The Dogfish are nocturnal and during the day are frequently seen 'asleep' on mud or sand. From 3-400 m. At night they become active and feed on any small animals they find living on the bottom. At the end of the summer they move into deep water for mating and then return to shallow water for spawning which occurs at different times according to locality. The eggs have yellowish-brown transparent cases 5-6.5 cm. long and 2-3 cm. wide with tendrils at the corners by which they are attached to seaweeds and other suitable objects. Day (1880) reports, that the female will be found wriggling round and round any suitable object until she has securely attached the tendrils. She then draws the 2 eggs from her body and the remaining tendrils can be attached also . The embryos take about 9 months to develop and can be seen clearly through the cases.

Edible, usually sold as rock salmon.

Large Spotted Dogfish; Nurse Hound; Bull Huss
p. 29
Scyliorhinus stellaris (L.)   **pl. 143**

Mediterranean; East Atlantic north to Norway, North Sea, English Channel.

Body slender and elongate gradually tapering towards tail; *snout* slightly more rounded than in the common dogfish; *eyes* oval, without nictating membranes but with a thick fold of skin on lower margin; *nostrils* present. Nasal flaps complicated and do not connect with the mouth as in the Common Dogfish (*Scyliorhinus canicula*) (see diagram); *gill slits*, 5 small the last two situated over the pectoral fins; *skin* rough; *fins* 2 dorsal fins the 2nd commences above

the middle of the anal fin and the 1st directly behind the pelvic fin. Tail fin only slightly directed upwards, the lower lobe larger than the upper.

Back greyish or brownish gradually becoming lighter on sides and nearly white underneath. The back and sides have circular brown spots with a light centre and many smaller black spots. The colour is not a good guide to these fishes as it can vary with age and locality.

Up to 1 m.

The Nurse Hound is less common than the Rough Hound and is usually found amongst rocks in quiet water. Most common from 20-60 m. They are nocturnal and feed by scavenging or on any animals that live on the bottom. The eggs are laid in cases which are dark brown 10-13 cm. long and 3.5 cm. wide with tendrils at each corner. They are usually laid attached to seaweed probably throughout the year.

The skin is very rough and was used for polishing wood, alabaster and copper under the name of 'rubskin'.

Black-mouthed Dogfish
p. 29
Galeus melastomus Rafinesque-Schmaltz
= *Pristiurus melanostomus* Müller & Henle

Mediterranean, East Atlantic north to Scandanavia south to Madeira

Body very slender and elongate; *snout* somewhat elongate; *eye* large, oval without nictating membranes; *gill slits*, 5, the 5th situated over the pectoral fin; *fins* 2 dorsal fins both small and similar in shape. 1st dorsal situated behind the pelvic . fin. Anal fin has an elongate base. Pectoral fin large. Tail fin long, more than 1/4 total body length, only slightly directed upwards with the lower lobe larger than the upper. Along the upper edge of the fin is a row of flat spines arranged like a saw.

Back brown to yellow, sides lighter, underneath whitish. Along the back and sides are irregular darker blotches more or less arranged in rows. The inside of the mouth is

black.
Up to 90 cm.

Found in deep muddy water, usually between 180-900 m. but occasionally as shallow as 55 m. They are carnivores feeding on bottom living animals but they also hunt fish in mid-water. Eggs are laid throughout the year in the Mediterranean and are distinct as the egg cases have no tendrils at the corners. They measure about 6 cm. long and 3 cm. wide.

Family:

TRIAKIDAE

The **Smooth Hounds** live in moderately shallow water. They have two dorsal fins of which the first is in front of the pelvic fins. The tail fin is not moon-shaped. There is no nictating membrane but the lower eyelid has a longitudinal fold. The teeth are small and rounded, with 3-4 points. Viviparous or ovoviviparous.

Smooth Hound
Mustelus mustelus (L.)

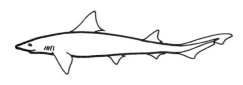

Lower lobe of tail fin larger than upper and directed upwards;
1st dorsal fin in front of pelvic fins;
colour uniform.
1.6 m.

Mediterranean; East Atlantic north to Britain.

Body slender; *snout* rather long and pointed; *eye* oval, without nictating membranes but with a thick fold of skin on lower margin; *teeth* mosaic-like; *gill slits*, 5, the 5th situated over the pectoral fin; *skin* rough, the scales oval with ridges which do not extend the full length of the scale; *fins* 2 dorsal fins, 1st situated behind the pectoral fin and the 2nd commences in front of the anal fin. Tail fin angled upwards and has a large notch.
Usually a uniform grey in colour with lighter sides. Occasionally there may be black spots scattered over the surface. Underneath whitish.
Up to 1.6 m.

Usually found on sandy and muddy bottoms from about 5 m. down to the Continental shelf. Mainly nocturnal they feed on any bottom living animals. Viviparous. A maximum of 28 are born in one litter.

Stellate Smooth Hound   pl. 144
Mustelus asterias Cloquet

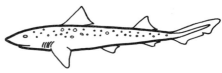

Lower lobe of tail fin larger than upper and directed upwards;
1st dorsal fin in front of pelvic fins;
white spots on back and sides.
2 m.

Mediterranean; East Atlantic north to Scotland south to Canaries; North Sea; English Channel.

Body slender; *snout* rather pointed; *eye* oval without nictating membranes but with

31

a thick fold of skin on lower margin; *nostrils* completely separate from mouth; *teeth* mosaic-like in 5-7 rows; *gill slits* 5, the 5th situated over the pectoral fins; *skin* rough, the scales rounded with one large central ridge and two smaller side ones which extend the full length of the scale; *fins* 2 dorsal fins, 1st starts immediately behind the pectoral fin base, 2nd dorsal starts in front of anal fin. Tail fin angled upwards, a distinct notch on rear edge.

Back light to dark grey; sides lighter, underneath whitish. Irregular white spots are scattered over the back and sides, particularly along the lateral line.
Up to 2 m.

Found on muddy sand from 5-165 m. Predominantly nocturnal they feed on any bottom living animals but principally crustaceans and occasionally fishes. Ovoviviparous. Up to 15 may be born in one litter.

Family:

CARCHARHINIDAE

Moderate to large **sharks** with 5 gill slits. The upper lobe of the tail is longer than the lower; the tail is therefore not 'half-moon' shaped. There are 2 dorsal fins of which the 1st is the largest. There is an anal fin. The eyes have a nictating membrane. The spiracles are very small or absent. The teeth are blade-like with only one point and only one or two rows are functional. Viviparous.

Blue Shark
Prionace glauca (L.)
= *Carcharhinus glaucus* Norman & Fraser

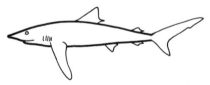

Very long sickle shaped pectoral fins;
long pointed snout;
colour blue.
4 m.

Mediterranean; all tropical and temperate seas.

Body long and slender with distinct upper and lower pits immediately in front of the tail fin; *snout* long and pointed; *spiracles* not present; *eyes* with nictating membranes; *teeth* (see diagram) pointed; *gill slits* 5, the

5th situated over the pectoral fin; *skin* with very small scales hence a smooth feel; *fins* 2 dorsal fins, 1st much longer than the 2nd and situated between the pectoral and pelvic fins. Tail fin large, upper lobe longer than lower and with a notch on lower surface. Pectoral fins very long and sickle shaped.
This shark is easily distinguished in life by its uniform dark blue back, lighter sides and white undersurface.
Up to 4 m., occasionally longer.

The Blue Shark is an extremely voracious, mainly nocturnal animal, which pursues and feeds on shoals of fish, other sharks, squids and occasionally man. Although usually found near the surface over deep water, they do, however, approach shore during the summer months, where, in S.W. England they are fished from the bottom. They may form more or less unisexual shoals. Young fish are born alive during the summer. They are about 60 cm. long and there may be more than 50 in a litter.

Tope
Galeorhinus galeus (L.)
= *Eugaleus galeus* Gill

Mediterranean; East Atlantic; North Sea; English Channel; all temperate and tropical seas.

Body slender; *snout* long and pointed; *eyes* oval with nictating membranes; *teeth* (see diagram) triangular and pointed; *gill slits* 5, the 5th situated over the pectoral fin; *fins* 2 dorsal fins, 1st situated between the pectoral and pelvic fins but nearer the pectoral than the pelvics. 2nd smaller than the 1st and slightly in front of the anal fin. Tail fin with a large notch on the lower edge of the upper lobe. Pectoral fin large.
Uniform dark grey on back, sides lighter, underneath whitish.
Up to 2 m.

Bottom living sharks usually found on gravel or sand from shallow water down to 400 m. but they may also swim near the surface. Fairly frequently encountered near coasts in summer either singly or in small shoals. They feed mainly on fish but also on other animals associated with the bottom. Young sharks are born alive during the summer and 52 embryos have been recorded from one female though more usually between 20-40 are born in one litter. The Tope is a common and popular game fish. The fins are used for shark fin soup.

Family:

SPHYRNIDAE

The **Hammerhead Sharks** resemble the Carcharhinidae except that the front portion of the head is flattened and is extended into one lobe on each side.

Common Hammerhead Shark   **pl. 145**
Sphyrna zygaena (L.)
= *Zygaena malleus* Valenciennes

Typical 'hammerhead' shape to head;
front of snout rounded.
4 m.

Mediterranean; Atlantic north to Biscay;
occasionally English Channel; all temperate
and tropical seas.

Head greatly flattened and extended into
two lobes to give the characteristic 'hammerhead' shape; *snout* smoothly rounded
(see diagram); *eyes* situated at extreme ends
of the head lobes and with nictating membranes; *teeth* triangular and pointed; *gill
slits* 5, the 4th and 5th situated over the pectoral fins; *fins* 2 dorsal fins the 1st immediately behind the pectoral fin, 2nd much
smaller than the 1st and opposite the anal
fin. Tail fin with a long upper lobe notched
underneath.

Slate-grey, brown-grey or greenish on the
back, sides and under surface lighter. The
tips and rear margins of the fins usually
darker.
More than 4 m.

Usually found in deep water from the surface down to 400 m., occasionally inshore.
They are excellent swimmers and feed mainly on fish but also crustaceans and cephalopods. There are confirmed reports that they
have attacked man. The young are born
alive at about 50 cm. and with between
37-39 in a litter.

Hammerhead Shark
Sphyrna tudes (Valenciennes)

Typical 'hammerhead' shape to head;
indentation in centre of snout.

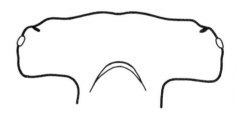

Mediterranean and all tropical oceans.

This shark is very similar to the Common
Hammerhead except that the centre of the
snout is indented and the body is rather
heavier.

34

Family:

SQUALIDAE

These are small sharks which chiefly live in very deep water. They are distinguished by having no anal fin and in most species there is a spine in front of each of the 2 dorsal fins. There are 5 gill slits, spiracles are present, the eye has no nictating membrane and the nostrils are separate from the mouth. There is a longitudinal fold of skin in the rear part of the body behind the pelvic fins. Viviparous.

Picked Dogfish; Spurdog; Common Spiny Dogfish
Squalus acanthias (L.)
= *Acanthias vulgaris* Risso

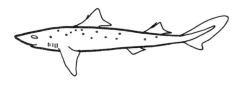

Spines in front of dorsal fins, 2nd dorsal spine not as long as the fin;
white spots on back and sides;
no anal fin.
1.2 m.

Mediterranean and Black Sea; East and West Atlantic; North Sea; English Channel.

Body slender; *snout* rounded; *eye* large and oval; *teeth* 24-28 in upper jaw, 22-24 in lower; *gill slits*, 5, all situated in front of the pectoral fins; *scales* small; *fins* 2 dorsal fins with spines in front of them. Both spines are shorter than the fins. No anal fin. Tail fin with large upper lobe and no notches. 1st dorsal fin situated between the pectoral and pelvic fins.
Light grey, dark grey or brownish on the back, lighter on flanks and white underneath. The back and sides have irregular white spots which may disappear with age. Up to 1.2 m.

These are small common migratory bottom living sharks which may be found from shallow water down to about 950 m. within a temperature range of 6°C-15°C. Usually in shoals but may occasionally be found singly. Young fish are born alive after a pregnancy of 18-22 months, the female approaching shore to give birth. They feed mainly on fish but also on crustacea and coelenterates and will attack shoals of fish that have been netted. In 1882 there was considerable concern that some fisheries would have to cease due to the damage caused by these fish. The spines are slightly poisonous. Edible; were once considered excellent smoked.

Blainville's Dogfish
Squalus fernandinus Molina
= *Acanthias blainvillei* Risso

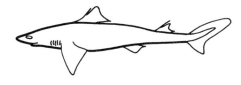

Spines in front of dorsal fins, 2nd dorsal spine at least as long as the fin;
no anal fin.
70 cm.

Mediterranean; East Atlantic from Gascony to South Africa; throughout all temperate oceans.

Body long and slender; *snout* rounded; *eye* oval and large; *gill slits* 5 all situated in front of the pectoral fins; *fins* 2 dorsal fins both with spines in front. The first dorsal spine is shorter than the fin but the 2nd dorsal spine is at least as long as the fin. The 1st dorsal is situated immediately behind the base of the pectoral fin. Tail fin with large upper lobe and no notch. Pectoral fins large. No anal fin.

Uniform grey or brown on the back, sides lighter, white underneath.
Up to 70 cm.

Occasionally seen in groups near the coast but more usually lives from 50 m. down to 700 m. Little is known about the habits of this fish except that it is carniverous.

Family:

SQUATINIDAE

The **Monkfish** are usually seen on the bottom where their flattened and elongated shape is unmistakeable. The pectoral fins are much enlarged. There are two dorsal fins which have no spines and there is no anal fin. There are 5 gill slits, large spiracles, large eyes with no nictating membrane and the nostrils are separate from the mouth. The nostrils have a valve which is generally fringed. The teeth are small and similar in both jaws. Ovoviviperous.

Monkfish; Angel Shark   p. 37
Squatina squatina (L.)   **pl. 146**

Mediterranean; East Atlantic from Scandanavia to the Canaries; English Channel.

Body flattened with a characteristic outline comprised of head, pectoral and pelvic fins and tail; *head* wide and rounded; *nostrils* with small slightly branched barbels; *eyes* small; *spiracles* large with diameter greater than eye diameter; *scales* small and rough, cover whole of upper surface and most of lower surface; *fins* pectoral fins very large, pelvic fins similar but smaller than the pectorals. No anal fin. 2 dorsal fins situated on tail behind the pelvic fins. Tail fin with lower lobe greater than upper.

Grey, brown or greenish on the back frequently with darker mottlings and occasionally with lines of lighter spots. Under surface white.
Up to 2 m.

By far the most common Monkfish. They live from 5-100 m. on sand and gravel and are usually seen lying on the bottom or partly buried beneath it. They feed on any animals which they find living on the bottom. The young are born alive during summer in Northern Europe and winter in the Mediterranean with 7-25 in a litter.

Monkfish
Squatina aculeata (L.)

Mediterranean and adjacent Atlantic.

Body flattened with a characteristic outline comprised of head, pectoral and pelvic fins and tail; *head* wide and rounded; *nostrils* with small very branched barbels; *eye* width equal to or less than spiracle width; *spiracles* with fringed front edge; *scales* small and rough, cover whole of back but underneath only on the front edge of the pectoral and

36

Monkfish *Squatina squatina* p. 36

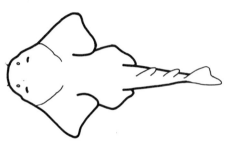

Rounded pectoral fins;
nostril barbels small and slightly branched.
2 m.

Monkfish *Squatina aculeata* p. 36

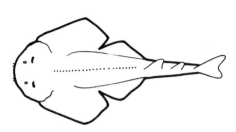

Row of large spines along centre back.
1.5 m.

Monkfish
Squatina oculata

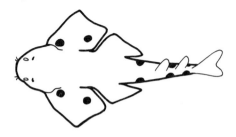

Pectoral and pelvic fins distinctly separate;
black spots on pelvic fins;
nostril barbels prominent and fringed.
1.5 m.

pelvic fins and along the centre of the tail. There is a distinct row of spines along the centre back and a number of spines between the eyes; *fins* pectoral fins large and rather rectangular. 2 dorsal fins, 1st situated immediately behind the pelvic fins.
Back sandy brown with symmetrically arranged white spots. Under surface white.
Up to 1.5 m.

Not common, usually found in muddy regions. Very little is known about this fish but its habits are probably similar to the Monkfish (*Squatina squatina*).

Monkfish
Squatina oculata Bonaparte

Mediterranean and adjacent Atlantic.

Body flattened with a characteristic outline comprised of head, pectoral and pelvic fins and tail; *head* wide and rounded; *nostrils* with small very branched barbels; *eyes* equal to or greater than spiracle width; *spiracles* with fringed front edge; *scales* rough and cover whole of upper surface but underneath only on the front edge of the pectoral and pelvic fins and along centre of tail; *fins* pectoral fins large and a distinct gap between them and the pelvics. 2 dorsal fins situated behind the pelvic fins.
Back reddish brown, brownish or grey with small white spots regularly distributed over the surface. There are also larger black spots on the pectoral fins and tail. Under surface white.
Up to 1.5 m.

Little is known about these fish except that they are usually found from 50-300 m. on mud or sand. Living fishes are their usual prey but they may only be able to catch them when the fish swim directly over their heads.

Order: HYPOTREMATA

Rays, Electric Rays and Guitar Fish.

The front margin of the pectoral fin is continuous with the head. The gill slits are on the lower surface of the head.

Family:

RHINOBATIDAE

The **Guitar Fish** are normally seen resting on the bottom and have a ray-like front half but are more shark-like behind. The pelvic fins are well separated from the pectorals. The two dorsal fins are about equal in length. The tail fin is well developed but there is no well defined lower lobe. Ovoviviperous.

Guitar Fish pl. 2
Rhinobatos rhinobatos (L.)

Characteristic shape.
1 m.

Mediterranean; East Atlantic north to Biscay.

Body, front half flattened and rear half shark-like; *head* is continuous with the body; *snout* elongated and pointed; *eyes* immediately in front of the spiracles; *skin* with a row of spines along the centre back; *fins*, pectoral fins continuous with head. 2 dorsal fins situated behind the pelvic fins. Tail fin not lobed.
Back brownish, greyish or yellowish. The dorsal and tail fins usually have whitish edges. Under surface white.
Up to 1 m.

Found in muddy and sandy areas either resting on the bottom or half buried. The 2 dorsal fins are a characteristic and conspicuous feature of this fish in life. Feeds on any bottom living animals. Young are born from February to July in the Mediterranean.

Family:

TORPEDINIDAE

The **Electric Rays** are able to deliver a shock of unpleasant intensity to a diver if touched. The head is a distinct disc with a separate rounded tail. There is a tail fin and two dorsal fins. The skin is naked. There is an electric organ on each side of the disc.
Electric Rays are usually encountered almost buried in the sand with only their spiracles and eyes visible. The outline of the disc is just discernable. They eat fishes which they envelop and stun with an electric discharge.

Eyed Electric Ray *Torpedo torpedo*

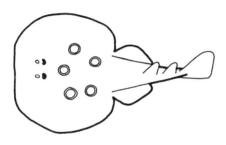

5 large spots on disc;
spiracles with no fringes or only slightly
fringed.
60 cm.

Marbled Electric Ray *Torpedo marmorata*

Marbled colouration;
6 to 8 small lobes on spiracle.
60 cm.

Dark Electric Ray *Torpedo nobiliana*   p. 40

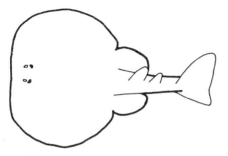

Uniform dark colour;
spiracles smooth.
130 cm.

Eyed Electric Ray   pl. 147
Torpedo torpedo (L.)
= *Torpedo ocellata* Rafinesque

Mediterranean; East Atlantic north to Biscay south to Angola.

Body disc shaped and contains 2 large electric organs; *tail* short and fat; *spiracles*
wider than eyes with a margin slightly
fringed or smooth; *skin* without scales; *fins*
pectoral fins large and form outer edge of
the disc.
Back light to dark brown with 5 (sometimes
1, 3 or 7) blue spots surrounded by black
and light brown rings. Under surface off
—white.
Up to 60 cm.

Found on or half buried in sand and
amongst sea grass beds. From shallow water to 50 m. occasionally down to 200 m.
They are solitary and nocturnal and feed on
fish and crustacea. The embryos take about
5 months to develop and the number of
young in a litter depends on the size of the
female.

Marbled Electric Ray   pl. 3 pl. 148
Torpedo marmorata Risso

Mediterranean; East Atlantic north to Brittany, south to South Africa; occasionally
English Channel.

Body disc-shaped and contains 2 large electric organs; *tail* short and fleshy; *spiracles*
with 6-8 small lobes; *skin* without scales;
fins pectoral fins large and fleshy and form
the outer margin of the disc.
Back light or dark brown with a darker
marbling. Under surface whitish with a
dark edge.
Up to 60 cm.

Found in shallow water usually from 2-20
m. occasionally down to 100 m. Usually
found on or half-buried in sand and mud.
They are solitary and nocturnal and feed on
bottom living animals including fish and

crustacea. In the Mediterranean the young are born in September and October with between 5 and 36 in a litter.

In the past the flesh of this fish was recommended to be eaten by epileptics and the electric organs to be applied to their heads for the therapeutic value of the shock. Underwater the shock gives a severe jolt but is not normally dangerous.

Dark Electric Ray p. 39
Torpedo nobiliana Bonaparte

Mediterranean; Atlantic north to Orkneys south to North Africa.

Body disc-shaped and contains two large electric organs. Young specimens have two dents on the front margin of the disc; *spira-cles* with smooth edges; *skin* without scales; *fins* pectoral fins large and form outer margin of the disc.

Back a uniform dark grey or black. Under surface whitish with dark edges.

Up to 180 cm.

Found on or half buried in sand and mud from 10-350 m. In the Mediterranean only occasionally above 100 m. They are solitary and nocturnal feeding mainly on fishes which they catch by stunning with an electric shock. A voltage as high as 220 v has been measured from this fish but underwater the voltage is probably much less. It is possible for a fisherman to receive a shock from his line when he has hooked a specimen.

Family:

RAJIDAE

The disc is diamond-shaped and the tail relatively small. The skin is spiny, the tail especially so. At the rear end of the tail are the two dorsal fins. Oviperous.

There is no biological distinction between **Skates** and **Rays** but those with long noses are normally called skates. Many spieces are biologically important as they are abundant on the trawling grounds and are good to eat. They are normally seen partly buried in the sand and are easy to overlook. They swim by flapping their pectoral fins in an undulating motion.

The Skates and Rays are very difficult to identify partly because there is great variation within the species and partly because the young differ from the adults and the males from the females.

Starry Ray *Raja asterias* p. 43

Yellowish spots outlined in dark dots.
Mediterranean.
70 cm.

Starry Ray *Raja radiata* p. 43

A central row of 12-19 very large spines in the adult;
northern distribution only.
90 cm.

Brown Ray *Raja miraletus*   p. 43

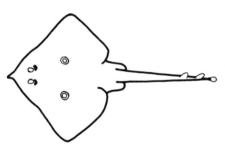

1 blue spot outlined in black and light brown on each wing;
prominent snout.
60 cm.

Blonde Ray *Raja brachyura*   p. 44

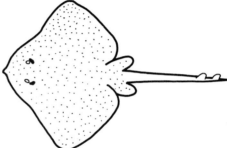

Small dark spots scattered over surface and extending to extreme edge of wings.
115 cm.

Spotted Ray *Raja montagui*   p. 44

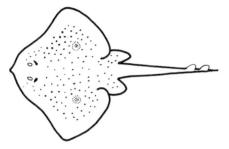

Small dark spots scattered over surface not extending to wing margins.
75 cm.

Undulate Ray *Raja undulata*   p. 44

Rounded pectoral fins;
wavy dark brown bands.
120 cm.

Painted Ray *Raja microcellata*   p. 44

Greyish or brownish with lighter spots and lines.
82 cm.

Thornback Ray *Raja clavata*   p. 45

Large broad-based spines scattered over surface.
85 cm.

Cuckoo Ray *Raja naevus*   p. 45

Large black and yellow marbled spot on each wing.
70 cm.

White Skate *Raja alba*   p. 46

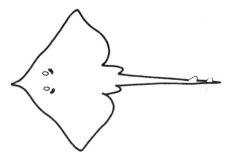

Elongate snout;
sharply angled wings;
underside white with darker edges.
200 cm.

Sandy Ray *Raja circularis*   p. 46

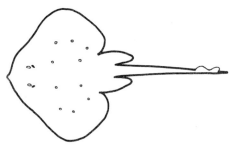

Rounded pectoral fins;
4-6 white spots on pectoral and pelvic fins.
120 cm.

Common Skate *Raja batis*   p. 47

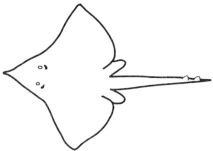

Elongate snout;
front edge of wings slightly concave;
grey underside.
235 cm.

Shagreen Ray *Raja fullonica*   p. 46

Snout slightly elongate;
no distinct markings.
110 cm.

Long-nosed Skate *Raja oxyrinchus*   p. 47

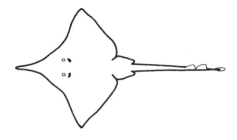

Very long snout;
grey or brown underside.
156 cm.

Starry Ray   p. 40
Raja asterias Delaroche
= *Raja punctata* Risso

Mediterranean.

Body disc shaped with a gently undulating
front edge; *snout* slightly pronounced but
blunt; *skin* rough on upper surface. Row of
spines (between 60-70) run from immediately behind the eyes to the 1st dorsal fin.
Two more rows of spines run along the tail.
Adult male fish have patches of spines on
the wings and at the sides of the head. There
are sometimes spines between the two dorsal fins.
Back brownish to brownish-red, olive-green
to yellow with a large number of small black
spots and fewer larger yellowish spots surrounded by a ring of small brownish-black
dots. The snout is light brown with no patterning.
Up to 70 cm.

Found on or half buried in mud and sandy
bottoms usually from 7-40 m. but also
down to 100 m. They feed on any bottom
living animals including fish and crustacea.
The egg capsules are a transparent greenish
brown, rectangular and measure about 45
mm. x 30 mm. and have elongated corners.
There are a number of rays in the Mediterranean with which this one may easily be
confused including, *R. brachyura* and *R.
montagui. Raja asterias* however, is far
more common than these species.

Starry Ray   p. 40
Raja radiata Donovan

North Sea; north East Atlantic; Iceland; also north West Atlantic.

Body disc shaped with rounded wings;
snout only very slightly pronounced; *skin*
rough. A central row of 12-19 large broad-
based spines run across the disc and along
the tail. There may be other large spines
either side of the central row. Under surface
smooth except on snout.

Upper surface light brown with many small
dark spots and fewer faint light cream spots.
Under surface white.
Up to 90 cm.
Found on mud, sand and stones from 20-
900 m. and usually within a temperature
range 1°C-10°C. They feed on bottom living animals including fish, crustacea,
worms and echinoderms. The egg capsules
are small and without horns. These fish are
common within their area and an important commercial species.

Brown Ray   p. 41
Raja miraletus L.
= *Raja quadrimaculata* Risso

Mediterranean; East Atlantic from Gascony to South Africa.

Body disc-shaped with rather pointed
wings; *tail* more than half total length of
body; *snout* elongate and pointed; *eyes*
larger than the spiracles; *skin*, in the adult
male spines are present along the front
edge of the disc and in small patches on the
wings. There is a central row of spines (between 14-18) along the tail which in the
male is flanked by one row of spines each
side and in the female by 2 rows of spines
each side. 2 spines are situated between the
dorsal fins and a number of spines around
the eyes; *fins* 2 dorsal fins and small tail fin.
Upper surface light brown with darker
edges and covered in small black dots. The
male fish also has yellowish spots. Snout region lighter. Each wing has a blue spot surrounded by a dark (nearly black) ring and
then a yellowish ring.
Up to 60 cm.

Found on or half buried in sand or mud.
Usually in water between 90-300 m. in
depth but in the summer they approach
shore and may be found as shallow as 30 m.
Egg capsules are laid in the Spring in the
Mediterranean and in winter in the Atlantic.

Spotted Ray p. 41 **pl. 149**
Raja montagui Fowler
= *Raja maculata* Montagu

Mediterranean; East Atlantic north to Shetlands south to Morocco; English Channel; southern North Sea.

Body disc shaped with wings angled; *snout* slightly elongate but blunt; *skin* spines only on front part of the disc. In adults they extend behind the eyes. There is one central row of spines which runs from behind the eyes along the tail to the 1st dorsal fin (20-30 spines in young, 40-50 in the adult). There may be 1-3 spines between the dorsal fins and adult males have small patches of spines on the wings.
Upper surface yellowish or brownish with small black spots which do not extend to the edge of the disc. Often a yellowish spot surrounded by a ring of black dots is present on each wing. Under surface white.
Up to 75 cm.

Found on muddy, sandy or sand and rock areas from 25-120 m. They feed on bottom living animals including fish and crustacea. Egg capsules are laid in Spring and early Summer in the Channel and the embryos take about 5 months to develop. The capsules measure 64-77 mm. by 37-46 mm. and have a short horn at each corner. One side of the capsule is smooth and the other covered with a network of fine fibres.

Painted Ray p. 41 **pl. 150**
Raja microcellata Montagu

East Atlantic north to Ireland and Cornwall south to Morocco.

Body disc shaped with angled wings; *snout* prominent; *eyes* small; *skin* only front half of surface rough. Central row of small spines along back and tail and one similar row each side along the tail.
Upper surface greyish to brownish with small white spots and wavy lines. Under surface white.
Up to 82 cm.

Found on shallow sandy bottoms. In some localities it may be very common while in other similar areas it may be unknown. Egg capsules are probably laid in Summer, they measure 8.7-9.5 cm. by 5.4-6.3 cm. and have 2 long slender horns at one end and 2 short curved horns at the other.

Blonde Ray p. 41
Raja brachyura Lafont
= *Raja blanda* Holt & Calderwood

East Atlantic north to Shetlands, south to Madeira; English Channel; occasionally Mediterranean.

Body disc shaped with angled wings; *snout* slightly elongate and rounded; *skin* adult fish are rough over the whole upper surface with a central row of small spines along the tail. Young fish are smooth and have a central row of spines along body and tail and there are 2 more complete rows of spines along the tail. Adult females have incomplete side rows along the tail. Adult males have patches of spines on the wings and at the sides of the head.
Upper surface light brown with numerous small dark dots which extend to the edge of the disc and larger irregular lighter patches. Under surface whitish. In young fish the extreme tip of the snout is black.
Up to 115 cm.

Found on muddy and sandy bottoms between about 40-100 m. Young fish are found in shallower water than the adults. They feed on bottom living animals which include fish, crustacea and molluscs. The egg capsules are large (115-143 mm. by 72-90 mm.) and are elongated into horns at each corner. Up to 30 may be laid at a time during the Spring and early Summer. The embryos take about 7 months to develop.

Undulate Ray; Painted Ray p. 41
Raja undulata Lacépède

Mediterranean; East Atlantic north to Britain; English Channel.

Body disc shaped with rounded wings; *snout* slightly elongate; *skin*, back rough except for the base of the tail and the pelvic fins. A row of irregularly spaced spines along the centre back and tail. Adult males have 1 row of spines each side of the central row along the tail while adult females have 2 rows. There are also a number of spines around the eyes.

Upper surface grey brown, brown or brownish yellow. Scattered over the surface are darker wavy lines each bordered with white dots. Snout pinkish grey. Under surface white.

Up to 120 cm.

Found offshore on sand or mud down to 200 m. They probably feed on bottom living animals like the other rays. Egg capsules are laid during late summer in the Atlantic and spring in the Mediterranean. The capsules are large and measure 82-90 mm. by 45-52 mm., they are reddish brown and have one side covered with fibres; horns at each corner.

Thornback Ray p. 41 **pl. 151**
Raja clavata L.

Mediterranean; East Atlantic north to Scandanavia; North Sea; English Channel; West Baltic; Black Sea.

Body disc-shaped with angled wings; *snout* slightly elongate; *skin* very rough on upper and lower surface. The upper surface has large-based spines symmetrically scattered over it. In the female these large spines are also present on the under surface. Young fish and females have a central row of spines down the body and along the tail. The male has only these spines down the tail.

Upper surface greyish or brownish with lighter and darker mottlings. The markings are much clearer in young fish and consist of small black spots and larger yellowish spots surrounded by brown. Under surface

white with a darker edge.
Up to 85 cm.

These rays are very common outside the Mediterranean. They are found on mud, sand, gravel, and rocky and sandy areas from 2-60 m. but also down to 500 m. They are nocturnal and feed on bottom living animals mainly crustacea but also fish, worms, molluscs and echinoderms. The breeding grounds of the Thornback Ray are inshore, the females preceeding the males to the grounds by several weeks. The egg capsules are laid in winter in shallow water and are often seen washed up on beaches and called 'Mermaids Purses'. They measure 6-9 cm. by 5-7 cm., are greenish-brown and have pointed horns extending from each corner. One side of the capsule is covered with fibres. The embryos take about 5 months to hatch and the young fish are frequently seen close inshore. The males reach maturity at about 7 years and the females at about 9 years.

A large specimen will give an electric shock if lifted up by its tail in air.

Cuckoo Ray p. 42
Raja naevus Müller & Henle **pl. 152**

Mediterranean; East Atlantic north to Scandanavia; North Sea; English Channel.

Body disc shaped with rounded wings; *snout* not prominent; *skin*, upper surface rough except for smooth patches on each wing. In the adult there are 4 rows of spines along the tail, the 2 central rows continue onto the disc. Young fish have a central row of tail spines. There may also be patches of spines at the sides of the head, on the wings and around the eyes. The under surface is smooth except for the front edge; *fins* 2 dorsal fins close together towards the end of the tail.

Upper surface-light brown or grey-brown with lighter patches. On each wing is a conspicuous large black and yellow marbled spot. Under surface white.

Up to 70 cm.

Found from 20-150 m. They feed on bottom living animals which include crustaceans, worms and fish. The egg capsules are a transparent brown and measure 6-7 cm. by 3.5 cm. There are pointed horns at each corner and one pair are at least twice the length of the egg case. Although little information is available on the habits of this fish it is of considerable commercial importance.

Sandy Ray p. 42
Raja circularis Couch

Mediterranean; East Atlantic north to Shetlands and Denmark; English Channel.

Body disc shaped with rounded wings; *snout* prominent; *skin* rough on upper surface, smooth underneath except along front edge. Central row of spines along tail present in young fish, the adults have 4 rows of spines along the tail; *fins* 2 dorsal fins set close together.
Upper surface light, dark or reddish-brown with 4-6 small whitish spots arranged symmetrically. Under surface white.
Up to 120 cm.

This is not a common fish and lives in rather deep water between 70-300 m. on sandy bottoms. Like the other rays it feeds on bottom living animals probably fish and crustacea. The amber coloured egg capsules are probably laid in Spring.

Shagreen Ray; Fullers Ray p. 42
Raja fullonica (L.)

East Atlantic south to Biscay; West Iceland; occasionally Mediterranean.

Body disc-shaped with angled wings; *head* and *snout* rather pointed; *eye* large; diameter equal to the space between the eyes; *skin*, upper surface rough with two rows of spines along the tail. There may also be patches of spines at the sides of the head, on the wings and around the eyes. No spines between the dorsal fins. Underneath smooth

except for the front edge of the disc, base of tail and tail.
Upper surface a uniform grey or brown with small darker dots. Under surface white.
Up to 110 cm.

Found from 35-350 m. on sandy areas but possibly on rather rocky ground also. They feed on bottom living animals possibly mainly fish but also crustacea. The amber-coloured egg capsules are laid in deep water and measure 88-99 mm. by 46-47 mm. with long horns at each corner. One pair of horns are longer than the capsule. They are a commercially important fish caught mainly by long line but also by trawl.

White Skate; Bottle Nosed Skate p. 42
Raja alba Lacépède
= *Raja marginata* Lacépède

Mediterranean; East Atlantic north to Ireland and Cornwall; occasionally English Channel.

Body disc shaped with pointed wings; *snout* elongate and pointed; *skin* rough in adults except for a smooth patch in the centre of the disc. Young are smooth. There is a central row of curved spines along the tail and one lateral row each side of it. Under surface prickly in the adults except for smooth patches on the wings. Young smooth except on the snout. There is one spine between the dorsal fins.
Upper surface greyish or bluish in the adult and reddish brown in the young. The whole surface is scattered with irregular lighter spots. Under surface white with a grey edge to the pectoral fins in the adult and with a black stripe across the rear edge of the pectorals and the pelvics.
Up to 200 cm.

Found on sand and mud from 40-200 m. The younger, smaller fish are usually in the more shallow water. Egg cases are laid in the spring and the embryos take about 15 months to develop. The capsules measure 16-19 cm. by 13-15 cm. with horns at each corner and with the larger horns flattened. The White Skate in the 17th century was a prized food fish of the French but is now of little commercial value.

Common Skate; Grey or **Blue Skate**   p. 42
Raja batis (L.)

East Atlantic north beyond Iceland and Norway; North Sea; English Channel; occasionally Mediterranean.

Body disc-shaped with a slightly concave front edge to the wings which are sharply angled; *snout* prominent; *skin* rough on whole of upper surface in the male, on the front region in the female but smooth in the young. There is a central row of spines along the tail, sometimes other rows are also present.
Upper surface greyish, brownish or greenish with lighter and darker spots scattered over the surface. Under surface greyish with black dots but becoming lighter with maturity.
Females up to 240 cm., males up to 205 cm.
Common bottom living fish found on sand and mud from 30-600 m. Young fish may be found from very shallow depths. They feed on bottom living animals, mainly fish and crustacea, but also they make forays from the sea bottom when they catch other species of fish and cephalopods. The egg capsules are yellow and measure 143-245 mm. by 77-145 mm. with horns at each corner. The longer horns have bunches of filaments at their ends. The capsules are laid during the Spring and Summer.
The Skate is an important commercial fish, but is apparently rarely seen by divers.

Long-nosed Skate   p. 42
Raja oxyrinchus (L.)

East Atlantic north to Trondheim south to Senegal; Mediterranean.

Body disc shaped with the front edge of the disc concave, wings pointed; *snout* very long and pointed; *skin*, upper surface rough only on snout and front edge of disc. Underneath rough over whole surface. Male fish have 3 rows of spines along the tail and the female fish 2 rows.
Upper surface greyish brown to chocolate brown. Young fish may have small white spots scattered over the upper surface. Under surface greyish or brownish with small black stripes and dots. Up to 160 cm.
Found on or half buried in sand or mud from 48-300 m. Young fish are usually found shallower than 100 m. and mature fish deeper. Feed on bottom living animals, mainly fish but also crustacea. Egg capsules are laid in the Spring and Summer. They measure 14 cm. by 11 cm. and are covered in yellow fibres. In deep Icelandic and Norwegian waters there is another species *Raja nidarosiensis* which is similar but may be distinguished by its shorter snout (less than 1/3 disc width).

Family:

DASYATIDAE

The disc of the **Sting Rays** is generally more rounded than that of the Rajidae. There are no dorsal fins and none of the fishes in our area have tail fins. There are almost always one or two spines near the base of the tail with a venom gland at the base and a groove along which the venom is ducted to the tip. The skin is smooth. Viviparous.
The Sting Ray may drive its venomous barb deep into a limb often causing a wound large enough to need stitching. There is intense pain quite out of proportion to the size of the wound. People rarely die from the sting but the pain may have a paralysing effect. The

wound should be washed in salt water and then plunged into water as hot as can possibly be tolerated. When the pain slackens any piece of barb still in the wound should be removed. Treatment is then the same as for any other wound.

Common Sting Ray   pl. 4 pl. 154
Dasyatis pastinaca (L.)
= *Trygon pastinaca* Cuv.

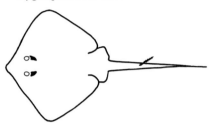

Tail 1 1/2 times disc length;
ridge above and below tail;
no spines except occasionally in the mid-line.
250 cm.

Mediterranean; Black Sea; East Atlantic south to Madeira, north to Norway; English Channel; occasionally West Baltic.

Body disc shaped with rounded angles to the wings and straight front edges. Body width only slightly greater than body length; *snout* fairly pointed; *tail* long about 1.5 times body length with one (sometimes more) large toothed spine situated about 1/3 the distance along the tail. There are ridges along both the upper and lower surface of the tail; *skin* smooth, old fish may have a central row of bony knobs along the mid-line of the body; *eyes* smaller than the spiracles which are directly behind them; *fins* no dorsal fins.
Upper surface grey, brown, reddish or olive-green. Young fish may have white spots. Under surface whitish with dark edges.
Up to 250 cm.

Sting Rays prefer calm shallow water (above 60 m.) where they can often be seen on, or half buried in, sand or mud. They can tolerate areas of low salinity and may be found in estuaries. The spine on the tail has a poison gland at its base and is capable of inflicting a very painful wound. Fish may be found with more than one spine as replacement spines develop before the old spine is lost. They are carnivorous and feed on any bottom living animals, fish, crustacea, molluscs etc. and can be very destructive in areas where shellfish are cultured. The sting ray is viviparous; the young are born in summer with between 6-9 in a litter. A number of legends have grown up around the Sting Ray. In one Circe is reputed to have given her son a spear with a Trygon's spine at the tip with which he later killed Ulysses.

Dasyatis centroura (Mitchill)

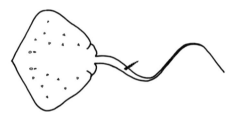

Tail 1 1/2-2 times as long as disc length;
scattered spines in adult;
ridge below tail.
3 m.

Mediterranean; East Atlantic north to Gascony; northern West Atlantic.

Body disc-shaped with rounded wings. The front edge of the wings are slightly, undulate. Disc width greater than length; *snout* small but pointed; *eyes* very small; spiracles large; *tail* at least twice as long as disc width with one or more poisonous spines situated towards the base. There is a well developed ridge on the underside of the tail; *skin*, there are a number of broad based spines scattered over the surface and in a row along the mid-line of the back. Young fish may not have any spines; *fins* 2 dorsal fins.

Upper surface yellowish to olive green. Lower surface whitish with a dark margin. More than 3 m. It is one of the largest Sting Rays in the world.

Not very common they are found on the bottom from 10-45 m., usually more common near the coasts in summer. Their habits are similar to the other Sting Rays.

Blue Sting Ray   **pl. 153**
Dasyatis violacea (Bonaparte)
= *Trygon violacea* Bonaparte

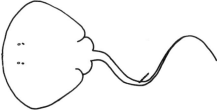

Gently curved front edge;
violet-grey or violet-brown on back, also on under surface but lighter.
1 m.

Family:

MYLIOBATIDAE

Mediterranean; warm water of East Atlantic.

Body disc shaped with curved wings. Front edge of disc curved; *snout* small, pointed; *tail* about twice the length of the body and very slender. In young fish the tail may be up to 3 times the body length. One or more poisonous spines are present; *skin*, there are a large number of small spines scattered along the mid line of the body; *fins*, no dorsal fins.
Upper surface greyish or brownish with a violet tinge. Under surface lighter.
Up to 1 m.

Found from shallow water to below 100 m. Although found on the bottom this fish leads a more active free swimming life than its near relatives. Feeds on small crustacea, fish, molluscs and cephalopods.

The **Eagle Rays** are more often seen actively swimming than are the other rays. The disc is diamond-shape and is wider than it is long. The tail is long and slender and at its base there are usually one or two venomous spines. The head is distinct forward from the level of the eye. The eye and spiracle are carried on the side of the head.

Eagle Ray *Myliobatis aquila*   p. 50   **Bull Ray** *Pteromylaeus bovinus*   p. 50

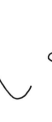 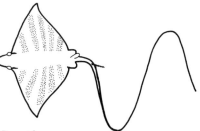

Semicircular head;   Snout longer;
uniform colour   dark diagonal striping on greenish background.
2 m.   2.6 m.

Eagle Ray   p. 49 **pl. 155**

Myliobatis aquila (L.)

Mediterranean; East Atlantic north to Scotland, occasionally Norway.

Body disc shaped. Body width nearly twice body length. The angles of the wings are pointed and have convex rear edges; *head* from eye level clearly distinct from the rest of the disc; snout rounded; *tail* twice as long as body. One or more spines about 1/4 the distance along the tail; *eyes* situated at the sides of the head; *teeth* 7 rows of large mosaic like teeth; *skin* smooth. Large fish may have small spines along the mid-line of the body; *fins* one small dorsal fin just in front of the tail spine.

Upper surface brown, grey, greenish or yellowish with a darker tail. Under surface whitish with darker edges to the pectoral fins.

Up to 2 m.

Eagle Rays are actively swimming fish which are often encountered near the surface and sometimes even jump out of the water. They are however, also found on sandy or muddy bottoms down to 300 m. They feed on bottom living animals, molluscs, crustacea etc. which are exposed by flapping the wings to disturb the sand or by digging with their snouts. 3-7 young are born in each litter.

Bull Ray   p. 49 **pl. 156**

Pteromylaeus bovinus (Geoffroy Saint Hilaire)

= *Myliobatis bovina* (Geoffrey Saint Hilaire)

Mediterranean

Body disc shaped, very pointed wings with convex rear edges; *tail* very long and slender about twice length of the disc. 1 spine situated towards the base of the tail; *head* separate from the body beyond the eyes; *snout* long and rounded; *eyes* situated at sides of the head; *spiracles* elongated and situated almost directly behind the eyes; *teeth* mosaic like in 7-9 rows; *fins* one small dorsal fin located immediately behind the pelvic fins.

Upper surface greenish brown with darker diagonal striping. Older fish are uniform in colour. Under surface white.

Up to 2.6 m.

This ray is found in deep cold or temperate water on mud or sand. In Spring they approach shore when they may be seen by divers. They feed on shellfish and fish and are a menace to the shellfish-culture industry. The young are born during late summer and early autumn.

Family:

MOBULIDAE

Resemble the Myliobatidae except that the two pectoral fins form two 'horns' at the front of the head. There are large gill slits. The food is plankton filtered from the water.

Devil Fish *Mobula mobular*

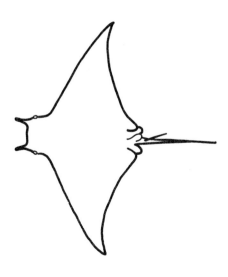

Distinctive lobes at each side of head.
6 m.

Mediterranean; East Atlantic north to Britain.

Body disc shaped; 2-3 times as wide as long. Wings pointed with convex front edges and concave rear edges; *head*, from the level of the eyes the head is distinct from the body. It is wide and has large 'horns' extending at each side of the head; *eyes* situated at each side of the head; *tail* slender with a spine near base; *teeth* minute in 150-160 series; *fins* small pelvic fins. 1 dorsal fin level with the pelvic fins.
Upper surface dark brown or black. Under surface white.
Up to 6 m.

Large fish which 'fly' through the water by gently flapping their very large pectoral fins. Usually in small groups occasionally near shore. Although extremely large these fish are harmless; they feed on small planktonic crustacea and fish which are funnelled into the mouth with the aid of the horns and filtered from the water by the gill apparatus. The young are born alive.

Class: **OSTEICHTHYES**

Sub Class: **ACTINOPTERYGII**

Order: **CHONDROSTEI**

Family:

ACIPENSERIDAE

The **Sturgeons** have asymetrical (heterocercal) tails, the body has several rows of bony plaques and the mouth is on the underside of the head.
Sturgeons are valuable fish since their eggs constitute caviar and their flesh is very good to eat.

Sturgeon
Acipenser sturio L.

Body covered with large bony plaques;
upper tail lobe longest;
mouth opening oval.
1.5 m.

North-East Atlantic, especially near the Rivers Gironde and Guadalquivir, Baltic, U.S.S.R., especially in lake Ladoga, Mediterranean especially in the Adriatic.

Snout elongate; *mouth* protrusible and situated on the underside of the head. Mouth opening oval; *barbels*, 4 between the snout and mouth; *armour* body covered with 5 rows of bony plaques (the row running down the centre of the flanks has 26-35 plaques), no scales; *fins*, the tail is asymetrical and shark-like, the spinal column being extended into the upper lobe of the tail.
Grey or green above, paler below, the fins are pinkish.
Sometimes attain 4 m. but females usually up to 1.5 m., the males smaller.
Generally live over sandy or muddy bottoms in brackish water. They feed on small animals and detritus which they dislodge by

rooting with their snouts and sucking them up through their extensible mouths. In spring they penetrate far up rivers to spawn. Males outnumber females and are smaller. The eggs stick to the bottom weed and stones. Caviar consists of the eggs taken from the female before they are laid. Sturgeon do not normally stray far from the rivers where they spawn but occasional specimens are taken which must have travelled hundreds of miles. A related species confined to the Adriatic *Acipenser maccari* Bonaparte has 33-42 plaques in the lateral row along the centre of the flanks and a shorter snout.

Huso huso (L.)
= *Acipenser huso* L.

Body covered with small bony plaques;
upper tail lobe longest;
mouth opening half-moon shaped.
6 m.

Eastern Mediterranean, Adriatic, Black Sea, Sea of Azof and Caspian Sea.

Resembles the sturgeon *Acipenser sturio* except that the snout is less elongate, the bony plaques are smaller and the mouth opening is half-moon shaped. Very large, may reach 6 m.

Super order: TELEOSTI

This order contains most of the present-day fishes. The vertebrae and skull are well-developed and bony.

Order: ISOSPONDYLI

One dorsal fin sometimes there is also a flap of tissue in the position of a second dorsal fin (adipose fin). The tail is symmetrical and forked and the swim bladder has an open duct.

Family: CLUPEIDAE

The **Herrings** and their relatives are long-bodied fishes generally living in shoals in open water. The body is covered with large easily detached scales. There is only one dorsal fin and no lateral line.
These are some of the most important of all food fishes.

Sprat p. 54
Clupea sprattus (L.)

North Mediterranean; east Atlantic from Norway to Gibraltar; North Sea; English Channel; Baltic.

Body oval in cross section but with a pronounced keel along the belly; *jaws*, lower longer than upper; *eyes*, the transparent (adipose) eyelids are present but are narrow; *scales*, there is a well developed saw-like keel of scales along the belly; *gill cover* ridged; *fins*, the dorsal fin begins midway between the hind edge of the eye and the base of the tail fin. The beginning of the pelvic fin is slightly in front of the beginning of the dorsal fin.
D 15-19; A 17-23.
Bluish above, silvery on sides and belly. A yellow-bronze band divides the blue back from the silvery flanks.
Up to 15 cm.

It is generally considered that there are 3 sub-species of Sprat, one in the Mediterranean, one in the north east Atlantic and one in the Baltic. They are able to tolerate water of extremely variable salinity from about 4 parts per thousand to full strength sea water of 36 parts per thousand. Sprats live in a temperature range of 4°-36°C. They spawn offshore from late winter to early summer preferring temperatures between 6° and 13° C. Unlike the herring the eggs float. In the winter large shoals come close inshore and in Norway they are captured in the fjords, canned, and sold as Brisling.

Herring p. 54 pl. 157
Clupea harengus (L.)

North East Atlantic from northern Europe north to the Arctic ocean. A smaller race lives in the Baltic.

Body oval in cross section, belly not keeled; *jaws*, lower slightly longer than the upper which is somewhat notched in the centre; *eye* there are transparent (adipose) eyelids; *gill cover* is smooth; *fins* dorsal fin begins midway between the snout and the base of the tail fin. The pelvic fin begins beneath the centre of the dorsal fin.
D. 17-20; A 16-18.
Dark blue or green above, silvery below.
Size variable, up to 40 cm.

Sprat *Clupea sprattus*   p. 53

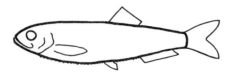

Single dorsal fin set back;
saw-like keel on belly.
15 cm.

Herring *Clupea harengus*   p. 53

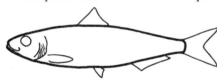

Single dorsal fin in middle of body;
no saw-like keel along belly;
gill cover smooth.
40 cm.

Pilchard (when large) **Sardine** (when small)
Sardina pilchardus   p. 55

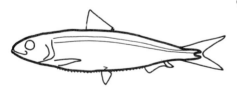

Single dorsal fin set slightly forward;
body not keeled;
gill cover ridged.
16 cm.

Gilt Sardine *Sardinella aurita*   p. 56

1 dorsal fin set forward;
belly keeled;
yellow stripe along flanks.
23 cm.

Allis Shad *Alosa alosa*   p. 56

1 or more spots on flank;
60-80 long thin gill rakers;
upper jaw notched.
40 cm.

Twait Shad *Alosa fallax*   p. 56

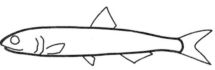

Usually 5-10 spots along flank;
20-28 stout gill rakers;
upper jaw notched.
50 cm.

Anchovy *Engraulis encrasicholus*   p. 56

1 dorsal fin;
lower jaw much shorter than upper.
20 cm.

The Herring is one of the most important food fishes of Northern Europe. From about the 12th to the 15th Century most Herring were landed through the Baltic ports but there was a sudden diminuation of stocks in the Baltic between 1416 and 1425. The North Sea Herring fishery chiefly operated by the Dutch took over and has remained important ever since. At the beginning of the 19th Century the Scottish Herring fleet began to develop and the British herring fishery reached a climax just before the 1914-1918 war; since then there has been a decline although it is still very important.

There are several races of Herring of which three are particularly distinct. The large Norwegian Herring can be 20 years old and becomes sexually mature at 5 to 8 years. Somewhat smaller are the fish from the southern North Sea that live for perhaps 11 years and mature at 3 or 4 years. The smallest race of Herring is the Baltic Herring or Strömling. These have fewer vertebrae than their Atlantic and North Sea cousins and they become mature when 2 or 3 years old.

There are about 50 spawning grounds for Herring in the North Sea alone and spawning occurs on each at a particular time of year from February to October. Once the eggs are laid they sink to the bottom where they lie in a carpet attracting fishes such as Haddock, which eat them. The newly hatched larvae live on planktonic plants, but as they grow larger they begin to feed on small free-swimming crustacea, sand eels etc. The very young herrings are long and slender resembling small sand eels from which they differ in their gut which extends nearly to the tip of the tail. At 5 cm. the herring takes on the proportions of the adult fish. These young fishes assemble in large shoals, often mixed with young sprats. Collectively they are known as whitebait and enter shallow coastal water and large estuaries. After spending about 6 months in the estuaries the herring appear to scatter into deeper water and do not join the main shoals until they are sexually mature.

Herring are not often seen by divers and this may be because they only come near the surface at night and generally keep near the bottom. They are either caught in drift nets at night or in trawls fishing over the bottom.

Pilchard (when large)
Sardine (when small)
Sardina pilchardus (Walbaum)

Mediterranean, east Atlantic from Canary Islands to S. Ireland, English Channel, rarely north to Norway.

Body rather rounded in cross section; *jaws* lower slightly longer than upper; *eyes*, there are transparent (adipose) eyelids; *gill cover* ridged; *scales*, the belly is smooth with no saw-like scaly keel; *fins*, the beginning of the dorsal fin is nearer to the snout than it is to the base of the tail; the pelvic fin begins below the centre of the dorsal fin. The last two rays of the pelvic fin are slightly elongate.
D 17-18; A 17-18.
Greenish or bluish above, silvery below. There may be a bluish band along the body.
Up to 16 cm., rarely 25 cm.

A very important food fish, especially in Spain and Portugal. They occur in very large shoals which may enter brackish water. They are found near coasts in the late spring and summer but in the autumn they disappear and probably over-winter in deeper water. Spawning in spring, summer and autumn close offshore but the time depends on the temperature. The eggs are free floating. The young of this fish are known as Sardines and are canned and exported from Spain and Portugal in large numbers. One of the most important methods of fishing is to attract a shoal with lights and then to surround it with a purse-seine net.
Divers tend to call any small silvery fish seen in shoals a sardine. The fish they normally see is a Sand Smelt (*Atherine* spp.) which differ from sardines in having two dorsal fins, also the Sand Smelt schools are much less mobile, the individual fish tend-

ing to hover in one place for a considerable period.

Gilt Sardine p. 54
Sardinella aurita Valencienne

Southern Mediterranean, both sides of the Atlantic, in the west from southern Spain to southern Africa.

Body oval in cross section, rather more compressed than the Pilchard (*Sardina pilchardus*) and with a pronounced saw-edged keel; *jaws*, lower jaw slightly longer than the upper; *eyes*, there are transparent (adipose) eyelids; *gill cover* without radiating ridges and with a single vertical one; *fins*, the beginning of the dorsal fin is nearer to the snout than to the base of the tail; the pelvic fin begins slightly behind the centre of the dorsal fin; the last two rays of the anal fin are slightly elongate; *scales*, there is a saw-like scaly keel along the belly; there are two large transparent scales on each side of the tail base, and one pointed scale towards the rear of the dorsal fin base.
D 17-20; A 16-18.
Bluish above, silver below with a longitudinal golden stripe along the flanks that quickly fades after death.
Up to 23 cm., rarely up to 38 cm.

A schooling species, it breeds in late summer near the coasts.

Allis Shad p. 54
Alosa alosa (L.)

Mediterranean, Biscay to northern Ireland, rarely further north, Baltic.

Jaw, lower jaw slightly prominent, mouth oblique; upper jaw notched to receive a knob in the centre of the lower jaw; *eyes* have transparent (adipose) eyelids: *gill cover* has radiating ridges; *fins*, dorsal fin begins nearer to the snout than to the base of the tail fin, the pelvic fin begins behind the beginning of the dorsal fin; *scales*, there

is a row of scales along the belly forming a saw-edged keel. Small scales cover the basal two-thirds of the tail fin; *gill rakers*, on the lower branch of the outer gill arch there are 60-80 long thin processes (gill rakers).
D 18-21; A 20-26.
Bluish above, silvery below. 1, or sometimes a succession of, dark shoulder spots.
40 cm., sometimes up to 60 cm.

Lives in deep water but in spring or early summer large shoals enter the rivers at night to breed. The eggs sink to the bottom. The young may spend up to 2 years in the rivers before going down to the sea, but the adults do not linger in the rivers after spawning and return immediately to the sea.

Twait Shad p. 54
Alosa fallax (Lacépède)
= *A. finta* (Cuvier)

Mediterranean and Eastern Atlantic from Biscay to Ireland. Rarely in more northern waters and Baltic.

Resembles the Allis Shad except that there are 20-28 stout gill rakers along the lower branch of the outer gill arch.
There is usually a large black blotch on the shoulder followed by a line of 5-10 smaller dark spots.
Up to 50 cm.

Very similar in habits to the Allis Shad. They spawn rather later in the year and do not travel so far upstream to breed. Several different races or sub-species are recognized. The Mediterranean race is known as *Alosa fallax nilotica* Geoffroy.

Anchovy p. 54
Engraulis encrasicholus Cuvier

Mediterranean and eastern Atlantic from Gibraltar no northern Norway.

Body slender and compressed, thicker along the back than along the belly which is not keeled; jaws, upper jaw much longer than

56

lower, the mouth cleft extends backwards to the hind edge of the eye; *eye*, there are no transparent (adipose) eyelids; *scales*, there are two large scales on the base of the tail fin. The scales do not form a saw-like ridge along the belly; *fins*, the dorsal fin begins midway between the snout and the base of the tail fin. Dorsal fin begins in front of the pelvic fin. D 15-18; A 20-26.

Greenish above, silvery below. In life a grey-blue band divides the dark upper part and the silvery lower part.

Up to 20 cm.

The habits are very similar to the Pilchard or Sardine (*Sardina pilchardus*). The anchovy is the subject of an important fishery in Portugal, Spain and Italy where they are canned or made into relishes and pastes. They are not eaten fresh.

Sub-order: SALMONOIDEI

Herring-like fishes which have a second dorsal fin composed of a simple flap of tissue without fin rays. Oviducts either absent or incomplete.

Family:

SALMONIDAE

Powerful small-scaled fishes which breed in fresh water. They are distinguished from related families by the vertebrae which are distinctly upturned at the base of the tail.

Salmon *Salmo salar*

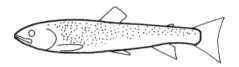

2nd dorsal fin without fin rays; upper jawbone scarcely reaches hind margin of eye.

Houting *Coregonus oxyrinchus*   p. 58

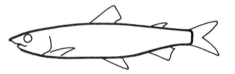

2nd dorsal fin without fin rays; jaw feeble; snout long and pointed.

Sea Trout *Salmo trutta*   p. 58

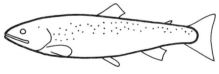

2nd dorsal fin without fin rays; upper jawbone reaches beyond eye. 140 cm.

Salmon   **pl. 5**
Salmo salar (L.)

East and West Atlantic, south to Biscay; Baltic.

Body, the tail stalk is narrower than in the Trout; *jaws*, the upper jaw bone extends back to the rear edge of the eye (in older males the jaws become grotesquely hooked in the breeding season); *scales*, there are 10-13 rows of scales between the 2nd dorsal fin and the lateral line; *fins*, there are two dor-

57

sal fins of which the 2nd is a simple flap of tissue without fin rays (adipose fin). The tail fin is shallowly forked. ID 12-14; 2D none; A 9-10.

Sea caught Salmon are steel-blue above, silvery below with rounded or x-shaped spots scattered on the upper part of the flanks and head. After the fish has entered fresh water the ground colour becomes greenish or brownish with streaks and spots of orange.

The young fish (parr) are bluish or brownish above with 8-15 broader dark transverse lobes extending onto the flanks with orange marks between each.

Spawning occurs in late autumn or winter in fresh water on gravel bottoms. After spawning the spent and exhausted females return at once to the sea, but the males may stay in fresh water until the end of the spawning season. The adult fish then spend from 5-18 months feeding in the sea before returning again to spawn. The young fish (smolt and parr) spend between 1 and 4 years in fresh water before entering the sea where they remain for at least a year. The older male parr may become sexually mature before entering the sea and often fertilise the eggs of the adult females.

The adult fish return to the waters where they were spawned being able to recognise the correct water by its particular smell. In the sea salmon feed on a variety of animals especially small fish and crustacea. It is a paradox that the adult salmon will take a bait in fresh water but when their stomachs are opened there is no evidence that they have swallowed food. Young salmon in fresh water take any suitable animal food.

Salmon are caught both by hook and line and by netting and are extremely valuable economically. The fish in pl. 5 was photographed whilst caught in a standing net on the north-east coast of England.

In 1965 a lucrative fishery for salmon was discovered off the Western coast of Greenland and it is believed that this is one of the feeding grounds of the European Salmon.

Sea Trout (marine); **Brown Trout** (fresh water) p. 57 **pl. 158**
Salmo trutta

Europe and North Africa, both fresh and salt water.

Body the tail stalk is broad; *jaw*, the upper jaw bone extends back well behind the rear edge of the eye, (in older male fish the lower jaw has a hook at the end but this is less pronounced than in the male Salmon); *scales*, small, 13-16 rows between the 2nd dorsal fin and the lateral line; *fins*, there are two dorsal fins, the 2nd being simply a flap of tissue without fin rays (adipose fin); the tail fin is square-cut or shallowly forked. ID 12-14; 2D none; A 10-12.

Sea Trout are dark grey above, silvery below with darker mostly x-shaped spots, especially on the gill cover; River dwelling brown trout are very variable in colour, greenish or brownish above, silvery below with black and red spots on the flanks and gill covers. There are no spots on the tail as there are in the Rainbow Trout *Salmo gairdneri*. The young fish (parr) are marked similarly to the Salmon but often have a white upper front edge to the dorsal fin.
Up to 140 cm.

The Brown Trout and the Sea Trout are the same species but the latter spends the greater part of its life in the sea. Spawning occurs in freshwater in late autumn and winter on shallow gravel bottoms. The young fish (smolts) may migrate down the rivers into the sea or remain in fresh water. Young trout feed chiefly on aquatic insect larvae and small animals that fall into the water. The adults feed also upon fishes and crustacea and winged insects. Sea Trout take a variety of small fishes and crustacea. It is probably the latter which gives the flesh of the Sea Trout its salmon-pink colour.

Houting p. 57
Coregonus oxyrinchus (L.)

Holland, Denmark, Germany and Sweden.

Body deep and laterally flattened; *head* small; *snout* long and conical; *jaws* small, the upper jaw is longer than the lower; the upper jaw bone reaches back to the 1st third of the eye; *teeth* in the jaws are feeble or absent; *fins*, there is a scaly flap of skin at the base of each pelvic fin. 2 dorsal fins, the second being a mere flap of tissue without fin rays. ID 12-15; 2D none; A 12-15.

Grey above, silver beneath; the snout and upper edge of the dorsal fin is dark grey or black.

In the North Sea the Houting lives in estuaries migrating up-river to spawn. In the Baltic there are several different forms living in lakes and rivers.

Family:

OSMERIDAE

Slender rounded bodies, slightly flattened along the sides. There is an adipose fin and the vertebrae are not upturned at the base of the tail.

Smelt; Sparling
Osmerus eperlanus (L.)

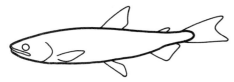

2nd dorsal fin without fin rays;
jaws extend beyond eye;
smells of cucumbers.
30 cm.

Baltic, North Sea, north-west coast of England and south-west Ireland; Atlantic coast of France. There is a related form in North America.

Body long and slender; *jaws* large extending back past the eye, the lower jaw is slightly longer than the upper; *odour* newly dead fish have a characteristic sweet smell resembling cucumbers, violets or rushes; *teeth*

present in both jaws (those in the lower jaw being the largest). There are also teeth on the roof of the mouth and on the tongue; *fins*, 2 dorsal fins of which the 2nd is merely a flap of tissue without fin rays (adipose fin). The pelvic fins are inserted below the first rays of the dorsal fin. ID 9-12; 2D none; A 12-16.
Greenish-grey above, silvery below with a wide silver band running along the flanks. The tail fin has a dark edge.
Up to 30 cm., sometimes more.

An inshore and estuarine fish which can survive in completely land-locked situations. They are voracious predators feeding on small fish and crustacea. They appear to assemble near river estuaries in winter and ascend the rivers to spawn in early spring.
Extremely good to eat when they are sufficiently fresh to retain their cucumber-like smell.

Family:

ARGENTINIDAE

Generally deep-water fishes, there is an adipose fin, vertebrae not upturned at the base of the tail, eyes and scales large.

Argentine
Argentina sphyraena L.

2nd dorsal fin without fin rays;
fresh fish almost transparent.

North-east Atlantic and Mediterranean. Not in southern North Sea or English Channel.

Body long and slender, the surfaces of the back, sides and abdomen are flattened; *jaws*, small not reaching back to the eye.

The upper jaw is slightly longer than the lower; *snout* sharply pointed; *teeth*, present on roof of mouth and on the tongue, none in the jaws; *fins*, two dorsal fins the 2nd being a simple flap of tissue without fin rays (adipose fin). The origin of the pelvic fin below the last ray of the 1st dorsal fin. ID 10-12; 2D none; A 11-15.
Yellow or greenish-grey above, silvery below with a silvery blue band along the flanks. The fresh fish is transparent enough for individual vertebrae to be seen.

Often trawled up from mixed sand and mud bottoms from 20-1000 m. Probably never seen by divers it feeds upon small bottom-living animals.

Family:

ESOCIDAE

There is only one genus in this family. The jaws are elongate and flattened. The teeth are erect in the lower jaw.

Pike pl. 33
Esox lucius L.

Fresh or brackish water;
long 'duck billed' snout;
dorsal fin set far back.
100 cm.

North temperate fresh and brackish waters of Europe, Asia and America, Baltic.

Body long and powerful almost equal in height throughout its length; *jaws* long and flattened giving a somewhat duck-billed appearance. The lower jaw is longer than the upper; *fins*, the single dorsal fin is set far

back above the anal fin. D 19-23; A 16-21.
Greenish or olive brown above, paler below. In the young fish there are lighter golden blotches and bands across the flanks. In the older fish there is a more irregular dappled pattern.
Up to 100 cm. or more.

Usually only found in fresh water where it is often seen by divers. It is common in the brackish waters of the Baltic. It is a voracious hunter and since it will take a trolled lure or live bait is a good sport fish. It is also very good to eat.
Spawning takes place in late winter or early spring in shallow water usually near the edge of marshy meadows. One or more males lie alongside the female who then darts away and it is at that moment that spawning occurs. The process may be repeated several times.

Order: ANGUILLIFORMES = APODES

The **Morays**, **Conger** and **freshwater eels** belong in this family. The body is long and typically eel-like, the skin has no scales or they are minute embedded in the skin. The dorsal, tail and anal fins form a continuous fringe. No pelvic fins.
The classification of this order is primarily based on skeletal characters.

Family:

ANGUILLIDAE

These **eels** always breed in the sea but may spend their juvenile life in fresh water. Pectoral fins are present.

Eel   **pl. 159**
Anguilla anguilla L.

Lower jaw longer than upper;
pectoral fin round;
eyes round, usually small.
140 cm.

Freshwater and neighbouring coasts of Europe, Asia and North Africa, Atlantic and Sargasso Sea. In North America there is a closely allied species *Anguilla rostrata*.

Jaws lower jaw longer than the upper; *eye* round, small in the yellow (immature) phase much larger in the silver (mature) phase; *scales* not visible but present embedded in the skin; *gill opening* a vertical slit in front of the pectoral fin; *fins*, the pectoral fin rounded; when depressed the pectoral fin terminates well in front of the origin of the dorsal fin.
In the immature yellow phase they are olive brown or greyish brown above, silver or yellowish silver below. In the more adult (silver) phase the back is dark grey-green and the flanks and belly are silver. (Colour differences are not, however, always sufficient to separate the two phases).
Up to 140 cm.

The yellow eel is the form normally seen or caught on hook-and-line. These fish have not yet reached sexual maturity. They are common in lakes and rivers and are often seen lying on the surface of the mud in deep lakes. Sometimes they stand with the upper part of their body upright in the water. Like the Conger and Moray they apparently spend some time in holes with only the head protruding. Sometimes seen in Estuaries and sheltered lagoons where they are often associated with clumps of kelp.
After a period of 7-20 years when the females measure between 50 and 100 cm. and the males 30-50 cm. a series of remarkable changes associated with the onset of maturity take place. The eye increases in diameter and becomes specialized for vision in deep ocean waters; the gut begins to atrophy and the gonads enlarge. The colour also changes from a yellow brown to a silver grey. In autumn the maturing eels set out on a long breeding migration. Usually on moonless nights they travel down the rivers to the sea (sometimes travelling short distances over land). These descending silver eels form the basis of a valuable fishery. Once they have left the coasts there is little further direct knowledge of their movements until their eggs are found in the deep waters of the Sargasso Sea (Western Atlantic). These eggs are found in spring and early summer; the young transparent blade-shaped larvae (the leptocephalus) gradually drift back to Europe and North Africa on

the prevailing current. The leptocephalus larva changes into the eel-shaped, but still transparent, glass eel during this passive migration. When the coastal waters are reached they assume the normal yellow brown colour of the young eel (elver). The entire drift across the Atlantic takes two or three years. When the elvers are about 5-10 cm. long they ascend the rivers, or more rarely settle in coastal waters. The passive migration of the larvae and the young eels from the Sargasso Sea to Europe and North Africa is proved beyond doubt, yet the opposite migration of the maturing eels is supported by no positive evidence. The yellow eel appears to be chiefly nocturnal and feeds upon bottom living animals including some fish.

Family:

MURAENIDAE

The **Morays** have no pectoral fins; the gill opening is small. The teeth are strong.

Moray Eel   pl. 7, 8
Muraena helena (L.)

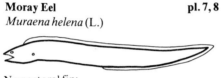

No pectoral fin;
gill opening small and round;
eyes round.
150 cm.

Mediterranean and warm Atlantic occasionally reaching as far north as Biscay.

Body long and powerful; *jaws* long, extending back beyond the eye; *teeth* long and sharp; *fins* no pectoral fin; *gill opening* is small and round; *eye* small and round.
Ground colour is usually dark brown mottled or marbled with yellow or whitish spots. The pattern is irregular at the front but tends to become more regular behind. A black spot surrounds the gill opening.
Up to 150 cm.

Occurs from near the surface to deep water. They are normally seen in deep cracks and crevices in the rock with just their head protruding, and seem to particularly favour old amphorae as homes.
Morays are caught rather more frequently during the winter when they approach the coasts to breed; the eggs are pelagic. They are extremely vicious when shot or hooked and they can inflict dangerous tearing wounds which often go septic. The flesh is good to eat, but it has been known since Galen that Moray flesh can be poisonous. The poison is of the 'Ciguatera type' and leads to stomach cramp and nervous disorders. The flesh probably becomes poisonous after the fish has eaten (perhaps indirectly) blue-green algae.

Family:

CONGRIDAE

The **Congers** have a gill slit; no scales and a pectoral fin.

Conger Eel   **pl. 6**
Conger conger (L.)

Pectoral fin pointed;
gill opening a long slit;
eyes elliptical.
200 cm.

Mediterranean; Atlantic, English Channel, North Sea and occasionally into the Baltic.

Body elongate and powerful; mouth large, extending back to the centre of the eye; *jaws*, upper jaw slightly longer than the lower; *scales* absent; *gill openings* large extending low onto the belly; *eye* elliptical; *fins* pectoral fin pointed with 17-20 rays. Dorsal fin origin only slightly behind pectoral.
Uniform grey-brown above, lighter below.
Up to 2 m., occasionally 3 m.

Lives on rocky ground and is particularly abundant in old wrecks. In the daytime the Conger lives in cracks and fissures with the head alone protruding. They may be found from a depth of a few meters to at least 1,000 m. Feed chiefly at night their diet being mainly bottom-living fishes, crustacea and cephalopods.
Congers probably spawn in mid-water over depths of 3-4,000 m.
There is one spawning area between Gibraltar and the Azores and this probably serves the northern European Conger population. There are also spawning areas in the Mediterranean. They have never been caught in the sexually mature state. Spawning occurs in mid-summer and the larvae drift back to the feeding grounds. The change from the larval to the adult form takes about 2 years and the young Congers tend to be found close inshore. The Conger spawns only once, and in the aquarium, at least, the teeth are shed and calcium is lost from the bones so that they become soft and gelatinous. The Conger is good to eat, especially in soups and once supported an important fishery.

Golden Balearic Eel   **pl. 194**
Congeromuraena balearica (Kaup)

Pectoral fin present;
fins with dark margins;
mouth small,
eyes round.
50 cm.

Mediterranean; Pacific and Atlantic north to Madeira.

Body elongate similar in shape to the Conger; *mouth* small, not extending behind the front margin of the eye. The upper jaw extends beyond the lower; *fins* pectoral fin present with 8-11 rays; *eye* round.
Greenish yellow or brown above, whitish below; the fins have a dark fringe.
Up to 50 cm.

Live on soft mud down to depths of 3,000 m. Little is known of their habits except that they are carniverous.

Order: SYNENTOGNATHI=BELONIFORMES

Elongated fish, sometimes extremely so. The lateral line runs low on the belly; dorsal fin set far back pelvic fin has 6 rays, no spiny rays. Swim bladder closed.

Family:

BELONIDAE

The **Garfish** and **Skippers** have very long bodies with very long and slender jaws.

Skipper; Saury Pike
Scomberesox saurus (Walbaum)

Elongate body and jaws;
finlets behind dorsal and anal fins.
50 cm.

Throughout the Atlantic and Mediterranean.

Body long; *jaws* elongate but of equal length except in fishes of 4-15 cm. when the lower jaw is the longest; *teeth* small; *fins*, the pectoral fin is directed downwards, finlets behind both the dorsal and anal fins. D 10-12 plus 5-6 finlets; A 12-14 plus 6-7 finlets.
Blue or olive above, golden or silvery below. A longitudinal silver band runs along each flank.
Up to 50 cm.

An oceanic species often seen in large shoals just beneath the surface. They appear to approach nearer the coasts in summer and early autumn but probably do not spawn inshore. The eggs are pelagic and are covered with filaments.
They have the habit of leaping out of the water and skipping along the surface to escape nets or predators.

Garfish; Garpike   pl. 160
Belone belone (L.)

Long body and jaws;
no finlets.
80 cm.

Mediterranean, North Atlantic and North Sea.

Body very long; *jaws* elongate, the lower being the longest; *teeth* large and pointed; *fins*, no finlets behind either the dorsal or anal fins. The dorsal fin is set slightly behind the anal. D 17-20; A 20-23.
Dark blue or dark green above with silver flanks and a yellowish tinge to the belly. The bones are green.
Up to 80 cm.

Primarily a shoaling fish over deep water but when the water is warm in summer or early autumn they may come close inshore, especially in the Mediterranean. Typically the swimmer sees them patrolling in small groups just beneath the surface. They are wary fishes and are difficult to approach.
Spawning is in coastal waters in summer. The eggs float and have adhesive filaments which become attached to floating weed etc. Garfish will take a moving bait and are thought to feed on small fishes (especially atherinids in the Mediterranean) cephalopods crustacea etc.

1. Lesser Spotted Dogfish *Scyliorhinus canicula* [H.L. Knook]. 2. Guitar Fish *Rhinobatos rhinobatos*
[D. Schofield].

3. Marbled Electric Ray *Torpedo marmorata* [W. Frei]. 4. Common Sting Ray *Dasyatis pastinaca* [H. Moosleitner].

5. Salmon *Salmo salar* [P. Carmichael]. 6. Conger Eel *Conger conger* [P. Smith].

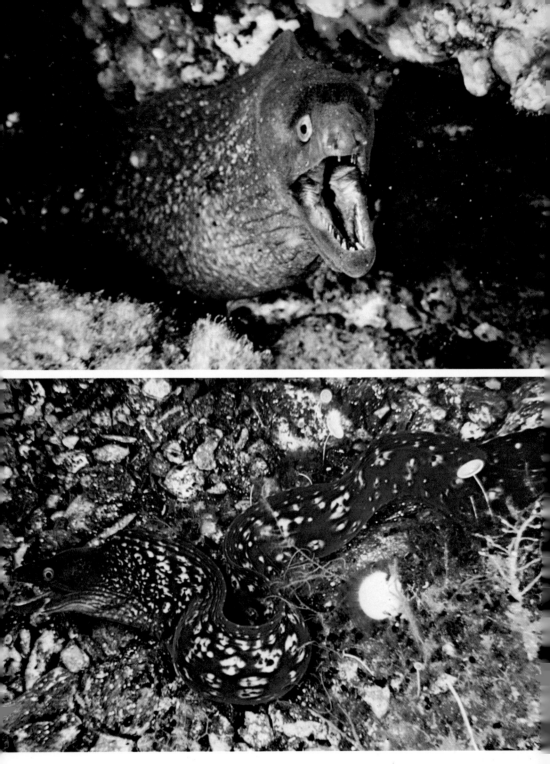

7. Moray Eel *Muraena helena* [R. Maltini, P. Solaini]. 8. Moray Eel *Muraena helena* [R. Maltini, P. Solaini].

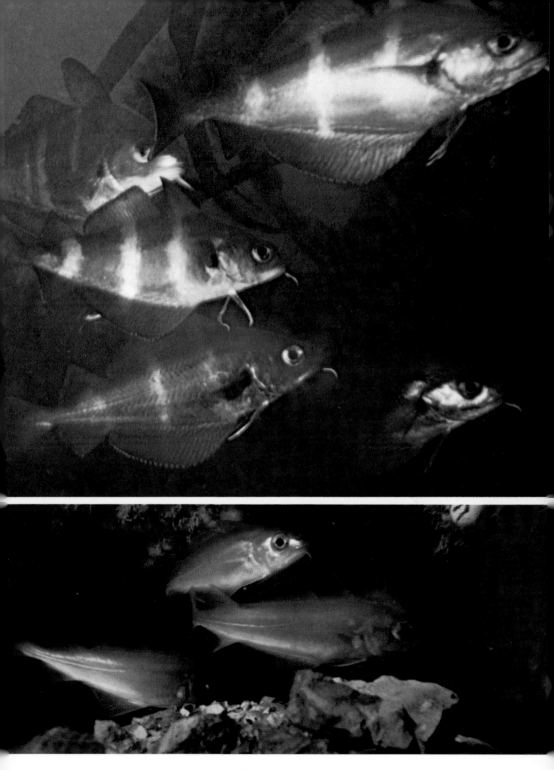

9. Bib *Trisopterus luscus* [K. McDonald]. 10. Poor Cod *Trisopterus minutus* [A. Baverstock].

11. Pollack *Pollachius pollachius* [K. McDonald]. 12. Saithe *Pollachius virens* [C.C. Hemmings].

13. Cod *Gadus morhua* [D.P. Wilson]. 14. Cod *Gadus morhua* (alternative colour form) [C.C. Hemmings].

15. Whiting *Gadus merlangus* [D.P. Wilson]. 16. Haddock *Melanogrammus aeglefinus* [C.C. Hemmings].

17. Ling *Molva molva* [A. Baverstock]. 18. Shore Rockling *Gaidropsarus mediterraneus* [G.W. Potts].

19. Three-bearded Rockling *Gaidropsarus vulgaris* [G.W. Potts]. 20. Great Pipefish *Syngnathus acus* [D. Clark].

21. John Dory *Zeus faber* [K. McDonald]. 22. Boar Fish *Capros aper* [D.P. Wilson].

23. Grey Mullet *Mugil* sp. [P. Carmichael]. 24. Cardinal Fish *Apogon imberbis* [P. Scoones].

25. Grouper *Epinephelus alexandrinus* [P. Scoones]. 26. Grouper *Epinephelus guaza* [P. Scoones].

27. Painted Comber *Serranus scriba* [P. Scoones]. 28. Painted Comber *Serranus scriba* [J. Lythgoe].
29. Comber *Serranus cabrilla* [W. Frei]. 30. Comber *Serranus cabrilla* [T. Glover].

31. *Anthias anthias* [N. Boeckler]. 32. Soldier Fish *Holocentrus ruber* [R. Maltini, P. Solaini].

33. Pike *Esox lucius* [J. Lythgoe]. 34. Dentex *Dentex dentex* [N. Sills].

35. Sea Bream *Pagrus pagrus* [J. Lythgoe]. 36. Marmora *Lithognathus mormyrus* [J. Lythgoe]. 37. Bogue *Boops boops* [J. Lythgoe].

38. Saupe *Boops salpa* [P. Scoones].

39. Black Bream *Spondyliosoma cantharus* (adult male) [G.W. Potts]. 40. Black Bream *Spondyliosoma cantharus* (young) [J. Lythgoe]. 41. Black Bream *Spondyliosoma cantharus* (young) [W. Frei].

42. Picarel *Maena chryselis* [C.C. Hemmings]. 43. Blotched Picarel *Maena maena* [N. Boeckler].

44. Red Mullet *Mullus surmuletus* [J. Lythgoe]. 45. Red Mullet *Mullus* sp. [A. Schuhmacher].
46. Goat Fish *Pseudupeneus barberinus* [C.C. Hemmings].

47. Brown Meagre *Sciaena umbra* [F. Knorr].

48. Horse Mackerel *Trachurus* sp. [P. Scoones]. 49. Horse Mackerel *Trachurus trachurus* (young) [P. Scoones]. 50. Horse Mackerel *Trachurus* sp. with jellyfish *Cotolorhiza tuberculata* [P. Scoones].

51. Amberjack *Seriola dumerilii* [Marineland]. 52. Pompano *Trachinotus glaucus* [D. Schofield].

53. Red Band Fish *Cepola rubescens* [D.P. Wilson]. 54. Sand Eel *Ammodytes* sp. [P. Scoones].

55. Damsel Fish *Chromis chromis* (adults in shoal) [J. Lythgoe].

56. Damsel Fish *Chromis chromis* (adults nesting) [J. Lythgoe]. 57. Damsel Fish *Chromis chromis* (young) [P. Scoones].

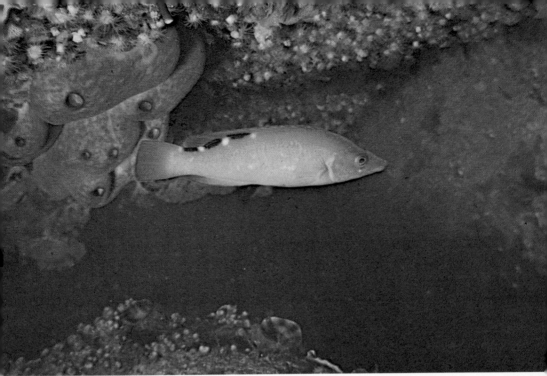

58. Cuckoo Wrasse *Labrus mixtus* (female) [T. Glover.] 59. Cuckoo Wrasse *Labrus mixtus* (male in breeding colours) [D.P. Wilson].

60. Ballan Wrasse *Labrus bergylta* [G.W. Potts]. 61. Ballan Wrasse *Labrus bergylta* (another colour form) [D.P. Wilson].

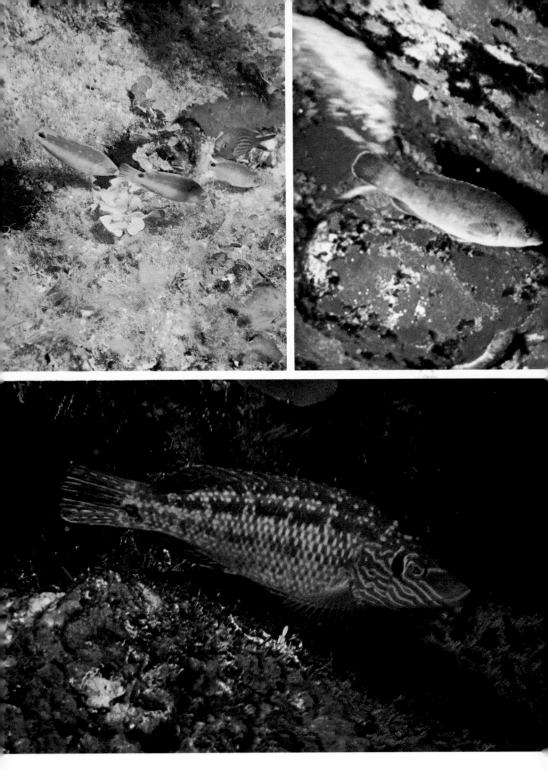

62. Axillary Wrasse *Crenilabrus mediterraneus* [J. Lythgoe]. 63. Brown Wrasse *Labrus merula* [R. Liepert]. 64. Corkwing Wrasse *Crenilabrus melops* (male in breeding colours) [P. Scoones]

65. Corkwing Wrasse (left) *Crenilabrus melops*. Rock Cock *Centrolabrus exoletus* [P. Scoones]. 66. Corkwing Wrasse *Crenilabrus melops* [P. Scoones]. 67. Rock Cock *Centrolabrus exoletus* (in breeding colours)[P. Smith].

68. Ocellated Wrasse *Crenilabrus ocellatus* (female) [P. Scoones]. 69. Ocellated Wrasse *Crenilabrus ocellatus* (male) [P. Scoones]. 70. Five-spotted Wrasse *Crenilabrus quinquemaculatus* [P. Scoones].

71. Painted Wrasse *Crenilabrus tinca* [J. Lythgoe]. 72. Painted Wrasse *Crenilabrus tinca* (alternative colour form) [H. Moosleitner]. 73. Painted Wrasse *Crenilabrus tinca* (alternative colour form) [N. Boeckler].

74. Cleaver Wrasse *Xyrichthys novacula* [A. Schuhmacher]. 75. Grey Wrasse *Crenilabrus cinereus* [R. Maltini, P. Solaini]. 76. Long-snouted Wrasse *Crenilabrus rostratus* [F. Knorr].

77. Rainbow Wrasse *Coris julis* (largest fish is male) [W. Frei].

78. Turkish Wrasse *Thalassoma pavo* (male) [A. Schuhmacher].

79. Turkish Wrasse *Thalassoma pavo* (female or immature left, male right) [J. Lythgoe]. 80. Turkish Wrasse *Thalassoma pavo* (intermediate colour form. There is a young grouper *Epinephelus guaza* behind) [P. Scoones].

81. Goldsinny (above) *Ctenolabrus rupestris* [A. Baverstock].
82. Lesser Weever *Trachinus vipera* [P. Scoones]. 83. Spotted Weever *Trachinus araneus* [H. Moosleitner].

84. Parrot Fish *Euscarus cretensis* (probably female) [J. Lythgoe].

85. Star Gazer *Uranoscopus scaber* [A. Schuhmacher]. 86. Star Gazer *Uranoscopus scaber* (in typical posture) [W. Frei]

87. Atlantic Mackerel *Scomber scombrus* [D.P. Wilson]. 88. Frigate Mackerel *Auxis thazard*
[J. Lythgoe].

89. Dragonet *Callionymus lyra* (female or immature) [G. Harwood]. 90. Dragonet *Callionymus lyra* (male, female behind) [D.P. Wilson].

91. Tompot Blenny *Blennius gattorugine* [P. Scoones]. 92. Tompot Blenny *Blennius gattorugine* (alternative colour form) [P. Scoones].

93. Horned Blenny *Blennius tentacularis* [N. Boeckler]. 94. Striped Blenny *Blennius rouxi* [H. Moosleitner]. 95. Red-speckled Blenny *Blennius sanguinolentus* [F. Knorr].

96. Shanny *Blennius pholis* [P. Scoones].

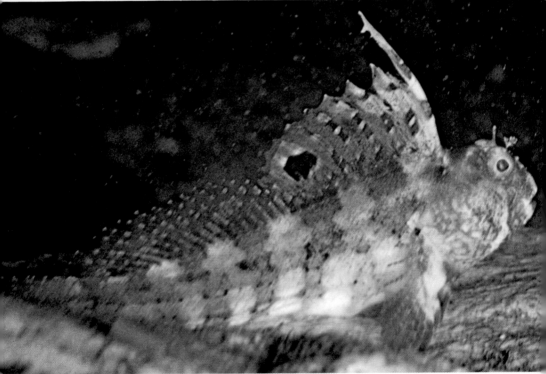

97. *Blennius sphinx* [P. Scoones]. 98. Butterfly Blenny *Blennius ocellaris* [D.P. Wilson]

99. Montagu's Blenny *Blennius galerita* [P. Miller]. 100. Yarrell's Blenny *Chirolophis ascanii* [D.P. Wilson].

101. Black-Faced Blenny *Tripterygion tripteronotus* (female) [P. Scoones]. 102. Black-Faced Blenny *Tripterygion tripteronotus* (male) [P. Scoones].

103. Black-Faced Blenny *Tripterygion tripteronotus* (male, alternative colour form) [P. Scoones].
104. Pigmy Black-Faced Blenny *Tripterygion minor* [P. Scoones].

105. Eel Pout *Zoarces viviparous* [H.L. Knook]. 106. Snake Blenny *Ophidion barbatum* [H. Moosleitner].

107. Two-Spotted Goby *Gobiusculus flavescens*. (in typical shoal) [A. Baverstock]. 108. Two-Spotted Goby *Gobiusculus flavescens* [C.C. Hemmings].

109. Black Goby *Gobius niger* [N. Boeckler]. 110. Black Goby *Gobius niger* [N. Boeckler]. 111. Rock Goby *Gobius paganellus* [P. Scoones].

112. Red-Mouthed Goby *Gobius cruentatus* [R. Maltini, P. Solaini]. 113. Slender Goby *Gobius geniporus* [R. Maltini, P. Solaini].

114. Leopard-spotted Goby *Thorogobius ephippiatus* [R. Maltini, P. Solaini]. 115. *Zosterissessor ophiocephalus* [H. Moosleitner].

116. *Gobius bucchichii* [P. Scoones]. 117. *Pomatoschistus pictus* [A.C. Wheeler]. 118. Sand Goby *Pomatoschistus (Gobius) minutus* [A. C. Wheeler]. 119. Giant Goby *Gobius cobitus* [P. Miller].

120. Gunnel *Pholis gunnellus* [C.C. Hemmings]. 121. Long Spined Sea Scorpion *Taurulus bubalis* [P. Scoones].

122. Scorpion Fish *Scorpaena scrofa* [G. Harwood]. 123. Scorpion Fish *Scorpaena scrofa* (alternative colour form) [J. Lythgoe].

124. Red Gurnard *Aspitrigla cuculus* [G. Harwood]. 125. Grey Gurnard *Eutrigla gurnardus* [D.P. Wilson.]

126. Streaked Gurnard *Trigloporus lastoviza* [D.P. Wilson]. 127. Saphirine Gurnard *Trigla lucerna* (young) [P. Scoones].

128. Flying Gurnard *Dactylopterus volitans* [J. Lythgoe]. 129. Flying Gurnard *Dactylopterus volitans* [J. Lythgoe].

130. Lump Sucker *Cyclopterus lumpus* [P. Krause]. 131. Lump Sucker *Cyclopterus lumpus* (male) Bib *Trisopterus luscus* in background. [J. MacMahon].

132. Cornish Sucker *Lepadogaster lepadogaster* [R. Maltini, P. Solaini]. 133. Connemara Sucker *Lepadogaster candollei* [R. Maltini, P. Solaini]. 134. Angler Fish *Lophias piscatorius*. [A. Baverstock].

135. Plaice *Pleuronectes platessa* [G. Harwood]. 136. Bloch's Topknot *Phrynorhombus regius* [H. Knook].

137. Turbot *Scophthalmus maximus* [G.W. Potts].

138. Sea Lamprey *Petromyzon marinus* [U.S. Fish and Wildlife Service]
139. Mako *Isurus oxyrhynchus* [J.N. Perez]
140. White Shark *Carcharodon carcharias* [Marineland of Florida]

141. Basking Shark *Cetorhinus maximus* (feeding) [C. Doeg]
142. Basking Shark *Cetorhinus maximus* (young) [J.N. Perez]

143. Large Spotted Dogfish *Scyliorhinus stellaris* [S.G. Giacomelli]
144. Stellate Smooth Hound *Mustelus asterias* [D.P. Wilson]

145. Common Hammerhead Shark *Sphyrna zygaena* [J.N. Perez]
146. Monkfish *Squatina squatina* [A. Schumacher]

147. Eyed Electric Ray *Torpedo torpedo* [S.G. Giacomelli]
148. Marbled Electric Ray *Torpedo marmorata* [J. Lythgoe]

149. Spotted Ray *Raja montagui* [D.P. Wilson]
150. Painted Ray *Raja microcellata* [D.P. Wilson]

151. Thornback Ray *Raja clavata* [J. James]
152. Cuckoo Ray *Raja naevus* [D.P. Wilson]

153. Blue Sting Ray *Dasyatis violacea* [S.G. Giacomelli]
154. Common Sting Ray *Dasyatis pastinaca* [A. Schuhmacher]

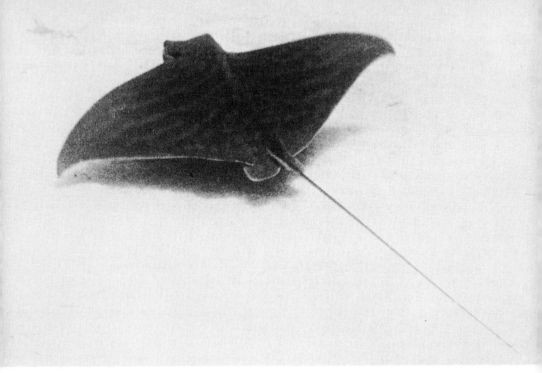

155. Eagle Ray *Myliobatis aquila* [J.N. Perez]
156. Bull Ray *Pteromylaeus bovinus* [H. Moosleitner]

157. Herring *Clupea harengus* [P. Svedsen]
158. Sea-trout *Salmo trutta* [D.P. Wilson]

159. Common Eel *Anguilla anguilla* [G. Harwood]
160. Garfish *Belone belone* [H. Fudge]

161. Bib *Trisopterus luscus* [C. Doeg]

162. Pollack *Pollachius pollachius* [C. Doeg]

163. Snake Pipefish *Entelurus aequorius* [G. Harwood]
164. Worm Pipefish *Nerophis lumbriciformis* [D.P. Wilson]

165. Great Pipefish *Syngnathus acus* [G. Harwood]
166. Deep-snouted Pipefish *Syngnathus typhle* [S.G. Giacomelli]

167. Sea Horse *Hippocampus ramulosus* [D.P. Wilson]

168. John Dory *Zeus faber* [C. Doeg]

169. Barracuda *Sphyraena sphyraena* [G. Lauckner]
170. Sand Smelt *Atherina sp.* [J. Lythgoe]

171. Bass *Dicentrarchus labrax* [C. Doeg]

172. Thick-lipped Grey Mullet *Mugil labrosus* [C. Doeg]
173. Wreck Fish *Polyprion americanum* [D.P. Wilson]

174. White Bream *Diplodus sargus* [A. Schuhmacher]
175. Sheepshead Bream *Puntazzo puntazzo* [F. Knorr]

176. Saddled Bream *Oblada melanura* [P. Scoones]

177. Two-banded Bream *Diplodus vulgaris* [P. Scoones]
178. *Diplodus cervinus* [G. Lauckner]

179. Red Sea Bream *Pagellus bogaraveo* [D.P. Wilson]
180. Marmora *Lithognathus mormyrus* [H. Moosleitner]

181. Blue Fish *Pomatomus saltator* [J.N. Perez]
182. Pilot Fish *Naucrates ductor* [L. Sillner]

183. Black-tailed Wrasse *Crenilabrus melanocercus* [G.W. Potts]
184. Parrot Fish *Euscarus cretensis* (probably male) [H. Moosleitner]

185. Streaked Weever *Trachinus radiatus* [J. Lythgoe]
186. Greater Weever *Trachinus draco* [A. Schuhmacher]

187. Skipjack Tuna *Euthynnus pelamis* (schooling) [U.S. Bureau of Commercial Fisheries, Biological
    Laboratory, Honolulu, Hawaii]
188. Skipjack Tuna *Euthynnus pelamis* [J.N. Perez]

189. Tunny *Thunnus thynnus* [J.N. Perez]

190. Sword Fish *Xiphias gladius* [J.N. Perez]
191. Dolphin Fish *Coryphaena hippurus* [Marineland of Florida]

192. Sand Eel *Ammodytes sp.* [D.P. Wilson]
193. Snake Blenny *Ophidion barbatum* [H. Moosleitner]
194. Golden Balearic Eel *Congeromuraena balearica* [H. Moosleitner]

195. Wolf Fish *Anarhichas lupus* [H. Hansen]

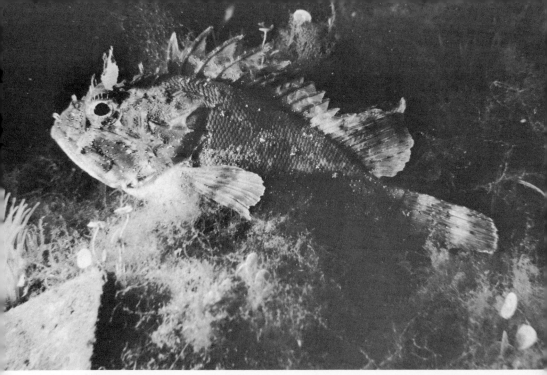

196. Scorpion Fish *Scorpaena porcus* [W. Frei]
197. Scorpion Fish *Scorpaena scrofa* [G. Harwood]

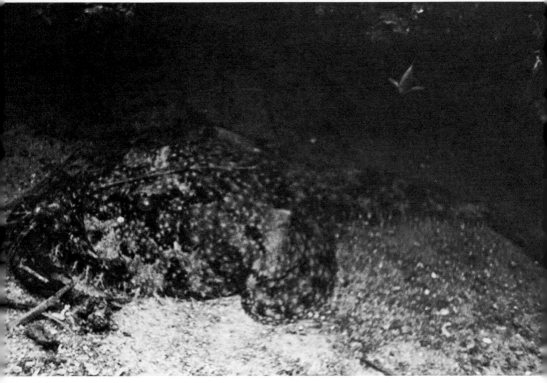

198. Fatherlasher *Myxocephalus scorpius* [G. Harwood]
199. Angler Fish *Lophias piscatorius* [G. Harwood]

200. Stickleback *Gasterosteus aculeatus* [H. Hansen]
201. 9-spined Stickleback *Pungitius pungitius* [A.C. Wheeler]
202. 15-spined Stickleback *Spinachia spinachia* [G. Harwood]

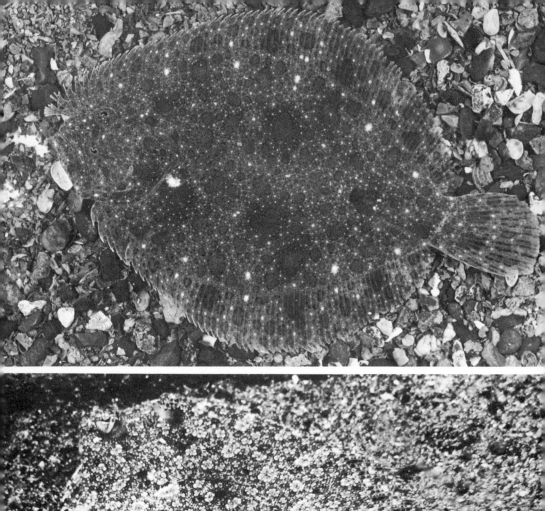

203. Brill *Scophthalmus rhombus* [D.P. Wilson]
204. Wide-eyed Flounder *Bothus podas* [A. Schuhmacher]

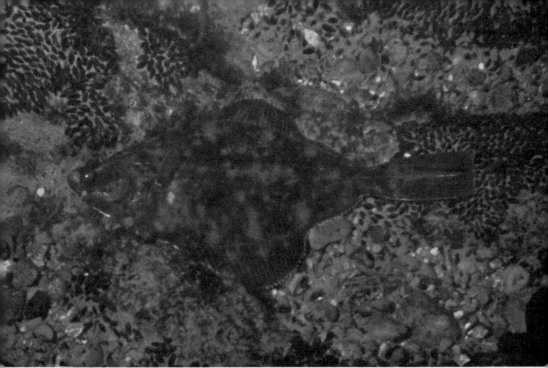

205. Flounder *Platichthys flesus* [A. Baverstock]
206. Dab *Limanda limanda* [D.P. Wilson]

207. Solenette *Solea lutea* [A. Baverstock]
208. Topknot *Zeugopterus punctatus* [G. Harwood]
209. Common Sole *Solea solea* [H. Knook]

210. Remora *Remora remora* [Marineland of Florida]
211. Trigger Fish *Balistes carolinensis* [S.G. Giacomelli]

212. Sunfish *Mola mola* [Marineland of Florida]

Family:

EXOCOETIDAE

The **Flying Fish** have very elongate pectoral fins and blunt snouts. The lower tail lobe is the longest. They swim just beneath the surface but when a boat approaches or predatory fish are near or, sometimes, for little apparent reason they fly over the surface of the water using the outspread pectorals as wings. The power for flight is given by the tail which beats in a rapid skulling motion and just before take off only the lower lobe is submerged. Flights of 200 m. or so have been observed, but frequently the fish falls back into the sea long before that. In a well-controlled flight the tail fin is the first to touch the water and it often imparts impetus for an extended flight without the body of the fish becoming submerged. The Flying Fish in our area fall into two well-defined groups. In one group (*Exocoetus*) the pelvic fin is short and the fishes live an oceanic existence; the eggs are pelagic. In the other group (*Cypsilurus* and *Danichthys*) the pelvic fins are long; they are not so oceanic in distribution and their eggs are probably laid associated with floating weed etc. or float at the surface.
It is difficult or impossible to distinguish most flying fishes when seen from the deck of a ship.

Flying Fish   p. 170
Exocoetus volitans L.

Tropical Atlantic, rarely in western Mediterranean.

Snout short, less than 1/5 the distance from snout tip to tail fin base; *scales*, 6 complete rows of scales between lateral line and dorsal fin base. A seventh row of small scales at the base of the dorsal fin can sometimes be distinguished; *teeth* none; *gill rakers*, 29-37; *fins*, pectoral fin longer than 2/3 the distance from snout to tail base; pelvic fins less than the distance from snout to tail base. D 14; A 12-14.
Pectoral fins ('wings') light grey without distinct markings; the rear margin has a clear narrow border.
Up to 18 cm.

The most common species in the open Atlantic, it is rarely found near coasts. In the summer when the water becomes warm it may enter the Mediterranean but apparently does not spawn there. Spawning takes place in the open ocean and the eggs are pelagic. Ripe females can be found at all times of the year.

Flying Fish   p. 170
Exocoetus obtusirostris Günther

Resembles *Exocoetus volitans* except that there are 7 complete rows of scales between the lateral line and the dorsal fin base and 24-29 gill rakers.

Flying Fish   p. 170
Cypsilurus heterurus (Rafinesque Schmaltz)

Atlantic (chiefly north-east) north from about 20°N to southern Norway. Western Mediterranean.

Scales, 30-38 scales in front of the dorsal fin, 7-9 rows of scales (usually 8) between dorsal fin base and lateral line, 47-49 vertical rows; *teeth* only jaw teeth present; *fins*, dorsal fin height not greater than 1/10th length from snout to tail base; second pectoral ray branched, pelvic fins 1/3-1/2 length from snout to tail base. D 12-14; A 8-10.
Pectorals grey without distinct cross-bands, narrow outer margin.
Up to 30 cm.

This is the most likely flying fish to be seen

Flying Fish *Exocoetus volitans*   p. 169   **Flying Fish** *Danichthys rondeletii*

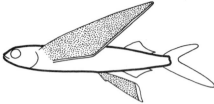

Pelvic fin short;
pectorals grey with clear border;
6 scale rows between dorsal fin base and
lateral line.
18 cm.

Pelvic fin long;
pectorals very dark.
23 cm. (Atlantic)
19 cm. (Mediterranean).

Flying Fish *Exocoetus obtusirostris*   p. 169

Pelvic fin short;
pectorals grey with clear border;
7 scale rows between dorsal fin base and
lateral line.
18 cm.

in the waters of N. W. Europe and possibly
the Mediterranean. Probably spawn in summer.

Flying Fish *Cypsilurus heterurus*   p. 169

Flying Fish
Cypsilurus lineatus (Cuv. et Val.)
= *Cypsilurus pinnatibarbatus* (Bennett)

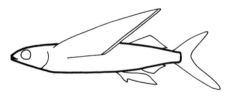

Warm waters of Atlantic from Cape Town to
Portugal, rarely in English Channel.

Scales 39-46 scales in front of the dorsal fin;
50-52 vertical rows of scales; *fins*, dorsal fin
height more than 1/10 the distance from
snout to tail base; second pectoral ray
branched; dorsal rays greater, rarely equal
to, the number of anal rays; length of pelvic
fin greater than 1/4 length from snout to
tail base. D 12-14; A 10-12.
Pectorals greyish, dorsal fin has a dark
mark.
30 cm., rarely up to 40 cm.

Pelvic fin long;
pectoral grey with clear narrow margin;
dorsal fin without spots.
Up to 30 cm.

Flying Fish *Cypsilurus lineatus*

The largest of the Atlantic flying fish. For
some reason it does not appear to penetrate
into the Mediterranean.

Flying Fish
Danichthys rondeletii (Cuv. et Val.)

Pelvic fin long;
pectoral grey with clear broad margin;
dorsal fin with black spot.
30 cm.

Atlantic from South Africa to Portugal,

perhaps in English Channel.

Scales 25-31 scales in front of dorsal fin, 7 complete transverse scale rows; *fins*, pectorals longer than 2/5 length from snout to tail base. Pelvic fin longer than 1/4 length from snout to tail base. Number of dorsal rays equal or less than number of anal rays.

Second pectoral ray not branched. D 10-11; A 11-13.

Pectoral fins very dark or black except the lowermost ray and outer margin which are not coloured; pelvic fins very dark or black; dorsal fin grey; anal fin clear. The rays of the tail fin are coloured.

23 cm. in Atlantic, 19 cm. in Mediterranean.

Order: ANACANTHINI

Tail fin formed mainly of dorsal and anal rays; closed swim bladder; no spiny rays.

Family:

GADIDAE

The Cod family contains such apparently diverse forms as the **Cod, Rocklings and Lings.** The first vertebrae are attached to the skull, the gill openings are wide and there is a separate tail fin.

An extremely important group of food fishes particularly in northern waters. Barbels are usually present and are a valuable aid in identification. Several cods have three dorsal fins and two anal which give a characteristic silhouette in the water. Other species are almost eel-like in shape and have one short and one very long dorsal fin.

Poutassou; Blue Whiting; Couch's Whiting
Gadus poutassou (Risso)   p. 172
= *Micromesisteus poutassou* Risso

North Atlantic, especially on the east coasts from north Norway to Biscay, deeper parts of North Sea and west Mediterranean.

Body long and narrow; *jaws*, lower slightly longer than upper; *barbel* absent; *lateral line* almost straight; *fins*, 3 dorsal, 2 anal, the interspaces between the fins are very wide especially between the 2nd and 3rd dorsals. The 1st anal fin originates beneath the first third of the dorsal fin and is rather elongated. The tail fin is slightly notched. 1D 12-14; 2D 12-14; 3D 23-26; 1A 33-39; 2A 24-27. Bluish-grey on back, flanks somewhat lighter. The belly is pale. The dorsal and anal fins are edged with black and the mouth cavity is black. There is often a diffuse black spot at the base of the pectoral

fin in the adult.

Up to 35 cm. in Mediterranean, 42 cm. in North Sea.

Unlikely to be met by divers and it has little commercial importance at present. Primarily a deep water pelagic species living in depths greater than 100 m. over water depths of up to several thousand metres. Young fish may occur in dense layers in the upper 10-30 m. but usually over bottom depths of 200 m. or more. They do not live in water colder than about 8°C. They spawn in mid water over considerable depths in winter and spring. The eggs are pelagic and the young, up to 15 cm., live near the surface at about 10 m. for a very long time during which they are distributed over a wide area.

Whiting   p. 172 **pl. 15**
Merlangius merlangus (L.)
= *Gadus merlangus* L.

171

Poutassou *Gadus poutassou*   p. 171

Space between 2nd and 3rd dorsal fins.
35 cm.

Whiting *Merlangius merlangus*   p. 171

Long upper jaw;
barbel minute or absent.
60 cm.

Bib *Trisopterus luscus*   p. 175

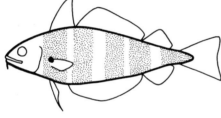

Usually vertical bands on body;
long barbel;
30 cm.

Poor Cod *Trisopterus minutus*   p. 175

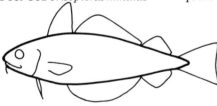

Uniform bronze colour;
barbel present;
20 cm.

Norway Pout *Trisopterus esmarkii*   p. 175

Barbel present;
lower jaw longest.
26 cm.

Pollack *Pollachius pollachius*   p. 176

Lower jaw longer than upper;
spaces between dorsal fins.
80 cm.

Saithe *Pollachius virens*   p. 176

Jaws almost equal in length;
space between dorsal fins.
100 cm.

Cod *Gadus morhua*   p. 177

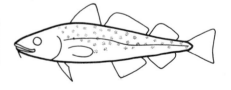

Light-coloured arched lateral line;
mottled body.
100 cm.

Haddock *Melanogrammus aeglefinus*

p. 177

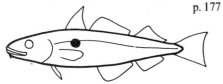

Black 'thumbprint';
black lateral line.
80 cm.

Hake *Merluccius merluccius*

p. 179

Long body;
long dorsal fin.
100 cm.

Tusk *Brosme brosme*

p. 178

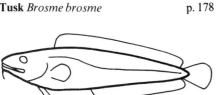

One dorsal fin;
dorsal and anal fins attached to the tail by
a membrane;
usually deep water.
60 cm.

Ling *Molva molva*

p. 179

Elongate body;
unforked barbel on chin;
upper jaw longer than lower.
100 cm.

Greater Fork-beard *Phycis blennoides* p. 178

Pelvic fin a single branched elongate ray;
elongate 3rd ray of 1st dorsal fin.
45 cm.

Mediterranean Ling *Molva elongata* p. 179

Elongate body;
forked barbel on chin;
lower jaw longer than upper;
dorsal, anal and tail fins with violet edges.
90 cm.

Lesser Fork-Beard *Phycis phycis*

p. 178

Pelvic fin a single branched elongate ray;
no elongate ray on 1st dorsal fin.
25 cm.

Tadpole Fish *Raniceps raninus*

p. 180

Tadpole-shaped;
1st dorsal fin has only 3 rays.
30 cm.

Shore Rockling   p. 180
Gaidropsarus mediterraneus

Three barbels;
not patterned.
50 cm.

Three-Bearded Rockling   p. 180
Gaidropsarus vulgaris

3 barbels;
bold dark pattern on red-brown background.
55 cm.

Four-Bearded Rockling   p. 180
Rhinonemus cimbrius

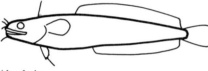

4 barbels;
deep water.
40 cm.

Five-Bearded Rockling *Ciliata mustela*

p. 181

5 barbels;
upper jaw not fringed;
shallow water.
25 cm.

Northern Rockling *Ciliata septentrionalis*

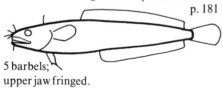

p. 181

5 barbels;
upper jaw fringed.

Eastern Atlantic from northern Norway to Biscay, North Sea, English Channel, Mediterranean, Faroes and Iceland. There is a subspecies in the Black Sea.

Jaws, upper longer than lower; *teeth* sharp and prominent; *barbel* very small or absent; *lateral line* gently curved; *fins*, 3 dorsal, 2 anal separated by short interspaces. The 1st anal fin is long and rounded and originates beneath the middle of the 1st dorsal fin. The tail fin is broad and square-cut. ID 12-15; 2D 18-25; 3D 19-22; 1A 30-35; 2A 21-23.
Body olive, sandy or bluish above, flanks silvery mottled or streaked with gold, belly white. The lateral line is golden bronze. The two anal fins are rimmed with white. There is also an inconspicuous diffuse dark spot at the base of the pectoral fin.
Up to 60 cm., sometimes 70 cm.

The Whiting has considerable commercial importance. It is predominantly a shallow water species and the bulk of the commercial catch is taken from 25-150 m. They are chiefly caught on sandy or muddy bottoms but venture further above the bottom than Haddock or Cod and are often caught near the surface. They are rarely seen by divers, even in water where they are known to abound. This may be because the neutral colouring of the fish when seen underwater blends in well with the background.
The food is mainly crustacea and fish. The proportions of fish in the diet increases with age. The Whiting appears to feed most actively at dawn and dusk.
There does not appear to be any well-defined spawning grounds and they breed throughout their geographical range. Spawning gets progressively later with increasing latitude. The peak is January in Biscay and May to June in Iceland. Spawning takes place at 5-10°C. A single individual has a prolonged spawning period of up to 8 weeks. The eggs and young are pelagic. The latter are often found in association with the jelly fish *Cyanea lamarcki* and *Chrysaura isoscoles*. This pelagic phase may last for more than a year. Whiting usually live for 5-7 years but occa-

sionally for 12 or 13 years.

Bib; Pout Whiting   p. 172
Trisopterus luscus (L.)   **pl. 9, 161**
= *Gadus luscus*

European Atlantic coasts and North Sea from Gothenburg to Biscay, English Channel, Western Mediterranean.

Body characteristically deep-bodied; *jaws* upper longer than lower; *barbel* prominant, equalling in length the diameter of the orbit, *lateral line* sharply curved; *fins* 3 dorsal, 2 anal with very short interspaces, the origin of the 1st anal fin is beneath or slightly behind the origin of the 1st dorsal fin. The 1st anal fin is deep and rounded and often joined to the 2nd anal. ID 11-14; 2D 20-24; 3D 18-20; 1A 30-34; 2A 19-22.
Flanks are pale copper coloured with 4-5 broad darker vertical bands which are sometimes very dark but occasionally very pale or absent (the bands are not seen in trawl-caught specimens since these have usually lost their scales). There is a small dark spot at the base of the pectoral fins. The margins of the anal fins are white and the margin of the tail fin is black. The lateral line is golden yellow but shows little contrast against the flanks.
Up to 30 cm., rarely 45 cm.

Of very little commercial importance but is often seen by divers in the English Channel. The young fish are found close inshore, often in large shoals. The larger fish move off into deeper water in depths of 30-100 m. and sometimes considerably deeper. They prefer a combination of rock and sand and the young fish, at least, have the reputation of always frequenting one place. Their food is chiefly molluscs and crustacea but the larger ones take small fish and cephalopods. The main spawning period is March-April but may continue until August. They prefer temperatures of 8-9°C and depths of 50-100 m.

Poor Cod   p. 172 **pl. 10**
Trisopterus minutus (L.)
= *Gadus minutus*

European shores of north Atlantic extending to Trondheim in the north, Kiel in the east, Faroes, round all British coasts. Western Mediterranean.

Body rather more slender and delicate than the Bib (*Trisopterus luscus*); *jaws*, upper longer than lower; *barbel* long and prominant; *lateral line* less curved than the Bib; *fins*, 3 dorsal, 2 anal fins set close together. The 1st anal fin begins between the middle and the end of the 1st dorsal fin. The 1st dorsal fin is rather pointed. 1D 13; 2D 23-26; 3D 22-24; 1A 28-29; 2A 23-25.
Uniform bronze-red above, flanks more coppery and belly a silvery grey. The lateral line does not show up well against the flanks. There is a small dark spot at the base of the pectoral fin.
Up to 20 cm., rarely up to 26 cm.

The Poor Cod has no commercial importance. It is found in the same sort of places as the Bib. Small individuals can be very numerous in shallow water, especially off rocky coasts. They appear to be attracted to artificial structures such as cages and framework on the sea bed. The food is chiefly bottom-living crustacea, molluscs and fish. They spawn in March and April, sometimes as late as July, usually in depths of 50-100 m. at temperatures higher than 8°C. The males live for 4 years, the females up to 6 years.

Norway Pout   p. 172
Trisopterus esmarkii (Nilson)
= *Gadus esmarkii*

North-eastern Atlantic from Iceland to south Ireland, common in North Sea, but rare in the southern part.

Body rather long; *barbel* present, but rather short; *lateral line* almost straight, *jaws*, lower extends beyond upper; *eye* large, its diameter exceeds the length of the snout; *fins*, 3 dorsal fins, 1st anal origin beneath the space separating 1st and 2nd dorsal fins. 1D 14-15; 2D 23-26; 3D 23-26; 1A 29-31; 2A 25-27.

Back bluish-olive, flanks silvery. There is a small black spot at the base of the pectoral fin.

Up to 26 cm. in Iceland, up to 20 cm. in the North Sea.

Often in very large shoals in depths of 100-200 m. at temperatures of 6-10°C. Most abundant over muddy bottoms.

Spawns in spring at temperatures of 5-10°C (optimum 6-7°C) at depths of 90-360 m. Eggs and larvae are pelagic. The young fish generally take up a bottom-living existence in August by which time they are 5-6 cm. long.

Pollack; Lythe   p. 172
Pollachius pollachius (L.)   **pl. 11, 162**
= *Gadus pollachius* L.

East Atlantic from North Cape of Norway to Biscay, S. E. Iceland, North Sea, northern parts of W. Mediterranean.

Jaws lower longer than upper, especially in larger individuals; *barbels* absent; *lateral line* commences above the gill cover, arches over the pectoral fin and follows the mid line from the origin of the 2nd dorsal fin to the tail; *fins*, 3 dorsal, 2 anal with well defined interspaces. The 1st anal is rather long and rounded and originates below the middle of the 1st dorsal fin. Tail fin broad and notched. 1D 12; 2D 19-20; 3D 17-19; 1A 29; 2A 17-20.

Coloration variable, the back is brown or olive, the flanks are paler, there are often dark yellow or orange spots or stripes scattered about the upper part of the body. The lateral line is greenish brown appearing dark against the flanks.

Up to 80 cm., may reach 120 cm. at an age of 15 years or more.

A very popular sport fish, often seen by divers, but with very little commercial importance. The young fish are often seen close inshore amongst rocks and weed where they may be confused with young Saithe (*Pollachius virens*). As the fish become older they move into deeper water, but it is not uncommon to see large specimens in shallow water either individually or in small groups.

The Pollack breeds in shallower water than the Saithe (usually shallower than 100 m.) and it prefers the slightly higher temperature of 8-10°C. The eggs and larvae are pelagic.

Saithe; Coalfish; Coley   p. 172
Pollachius virens (L.)   **pl. 12**
= *Gadus virens* (L.)

Both sides of north Atlantic. In west: Chesapeake Bay to Hudson Straight. In east: Greenland, Iceland, Barent Sea; British coast and south to Biscay.

Body streamlined and looks very symmetrical; *jaws*, in young individuals the jaws are equal in length, in large fish the lower jaw may be longer; *barbel*, so small as to be insignificant; *lateral line* almost straight; *fins*, 3 dorsal fins, 2 anal with clear interspaces. Beginning of 1st anal fin directly below the beginning of the 2nd dorsal fin. Tail fin broad with notch in rear margin. 1D 13-14; 2D 20-22; 3D 20-24; 1A 25-28; 2A 19-25.

Dark greenish brown or blackish—older fish are very dark. The belly is silvery white. The white lateral line shows up very clearly against the dusky flanks.

Up to 100 cm., sometimes up to 120 cm.

The Saithe is an important commercial species which is frequently seen by divers. It may be confused with the Pollack but its shape and colour together with the colour of the lateral line may allow a positive identification.

The young fish are found in shallow water close inshore amongst rocks and weeds. They are common in harbours in the North Sea. Older individuals move off into deep water where it is believed they undertake considerable migrations. Whilst in inshore waters the young feed on small fish and crustacea. The large, deep water, animals feed almost entirely on fish.

Spawning takes place in the early part of the year in rather deep water (100-200 m. in the North Sea). The eggs and larvae are pelagic, the young fish being carried inshore by surface currents. The optimum temperature for spawning appears to be 8-9°C but may occur between 3°C and 10°C.

large fish: **Cod**   p. 172 **pl. 13, 14**
smaller fish: **Codling**
Gadus morhua L.
- *Gadus callarias* L.

Both sides of north Atlantic from Cape Hatteras to Newfoundland. From Biscay to northern Norway; North Sea. There is also a race in the Baltic.

Jaws, upper longer than lower; *barbel*, long, almost equalling the diameter of the eye; lateral line is strongly curved to beneath the leading edge of the 3rd dorsal fin, and from thence runs straight to the tail fin; *fins*, 3 dorsal, 2 anal, all fins are rounded with only short interspaces. The first anal fin originates below or slightly behind the origin of the second dorsal fin. The tail fin is square cut. 1D 14-15; 2D 18-22; 3D 17-20; 1A 19-23; 2A 17-19.

Colour is very varied. The body is usually greenish-grey spotted with brown or grey. Some specimens taken from weedy rocky surroundings may be golden or even deep red in colour. The belly is usually a dirty greyish white. Young fish (5-10 cm.) have a distinct 'checkerboard' pattern on upper surfaces and flanks. The lateral line shows up conspicuously light against the ground colour of the flanks and in the sea the lateral line with its characteristic curve may be the only visible part of the fish.

Up to 100 cm., one cod of 169 cm. has been recorded.

The Cod is the most important commercial fish in its family. The young are often seen by divers either resting or browsing on the bottom. These young fish may be found in very shallow water close inshore but adults are found at all depths down to 500-600 m. Over its range it is found in temperatures of —4°C to +16°C but they seem to prefer water of 4-7°C.

The Cod is basically a bottom-feeder taking worms, crustacea and molluscs. However, considerable quantities of fish are taken, especially by the larger fish.

The Cod usually spawns on one of the well-defined grounds around 200 m. In the North Sea spawning may be as shallow as 20-100 m. at temperatures of 4-10°C. The Cod spawns earlier than the Haddock (*Melanogrammus aeglefinus*) and Whiting (*Merlangius merlangus*) in the northern North Sea, the peak being in February and early March. Courtship is well developed. In the aquarium the male erects all his median fins flaunting himself with much twisting and turning in front of the female. Courtship also seems to be accompanied by grunting or croaking noises. The eggs and larvae are pelagic and the young take up a bottom-living existence in the autumn of the first year when they measure 5-10 cm. In the North Sea they usually live for 6-10 years and reach about 80 cm. They may however live for much longer. A Cod caught in the Barents sea at 169 cm. was 24 years old.

Haddock   p. 173 **pl. 16**
Melanogrammus aeglefinus (L.)
= *Gadus aeglefinus*

Both sides of north Atlantic from Cape Hatteras to Newfoundland in the west, Spitzbergen to southern Ireland in the east. Also Faroe and Iceland. Throughout the North Sea but infrequent in the southern part. Sometimes in the eastern English Channel.

Jaws, upper longer than lower; mouth small; barbel short; lateral line gently curved; fins 3 dorsal, 2 anal. The first dorsal is high and sharply pointed all the fins are separated by distinct interspaces. The origin of the 1st anal fin beneath the origin of the first dorsal fin. The tail fin is slightly notched. 1D 15-16; 2D 19-21; 3D 19-22; 1A 23-24; 2A 22-23.

Dark purple to charcoal on the upper surface, sometimes olive. The flanks are dark silver and the belly is pale. The Haddock is easily distinguished by the large black 'thumbprint' behind the pectoral fin and by the black lateral line. In young fish the thumbprint may be ringed with white and is very conspicuous underwater.

Up to 80 cm., occasionally 100 cm.

A very important commercial species which is eaten either fresh or smoked. Although abundant in many areas they are seldom seen by divers and it is possible that they are frightened by the sound of conventional aqualung equipment (The Haddock in pl. 16 was lured into range by the broken up mussels (*Mytilus edulis*) spread on the bottom).

The Haddock usually lives very near the bottom but large shoals are occasionally found in mid-water. They chiefly feed on worms, molluscs, echinoderms but at times may gorge themselves on sand eels, Capelin and Herring spawn. Feeds at depths of 20-300 m.

Spawning takes place from 60-200 m. at temperatures of 6-10°C in the North Sea from late February to early May. There is a well developed courtship display involving both sound and sight signals. The eggs and larvae are pelagic. The young fish take up a bottom-living existence on the autumn of their first year when they are 7-15 cm. In some years the spawning is very successful and that year class can be caught in large numbers outside the normal range.

They usually reach an age of 6-8 years in the North Sea, 13 years in the Barents Sea and even older in the N. W. Atlantic.

Tusk p. 173
Brosme brosme (Müller)

North Atlantic, New Jersey to Newfoundland, Labrador, Greenland, Iceland, Barents Sea, northern North Sea and west coast of Britain south to North Ireland.

Jaws almost equal in length; *barbels*, a single barbel present; *fins*, one dorsal, one anal continuous through the tail fin to which both are connected by a thin membrane.

Brownish grey on back, somewhat paler below. Median fins have narrow white or pale yellow margins with a black band running immediately below them.

Up to 60 cm., exceptionally over 100 cm.

Unlikely to be seen by divers, they have some commercial value and are chiefly caught on long line. Sedentary bottom living fishes usually found singly or in small groups between 100 and 1000 m. They chiefly feed on molluscs and crustacea.

Spawning takes place from 50-400 m. (chiefly at 200 m.) in spring and early summer when the water is 6-9°C. An average sized fish may produce 2,000,000 eggs.

Greater Forkbeard p. 173
Phycis blennoides (Brünnich)

Adriatic, west Mediterranean, east Atlantic north to Scotland. Rarely North Sea and north to Norway and Greenland.

Barbel on chin; *fins*, 2 dorsal and 1 anal. 3rd ray of 1st dorsal is elongated being at least the height of the 2nd dorsal. The 2nd dorsal is long and of uniform height. Pelvic fins consist of a simple branched ray reaching back well beyond the vent. 1D 8-9; 2D 60-64.

Brown or grey, the dorsal, anal and tail fins yellowish with black edges. The young have 2 or 3 black blotches on the back.

Up to 45 cm., occasionally 75 cm.

Usually on sandy or muddy bottoms from 150-300 m. but may be caught as shallow as 10 m. They feed chiefly on crustacea but fish are sometimes eaten. They spawn from January to May in the Mediterranean and rather later in the northern waters.

Lesser Forkbeard p. 173
Phycis phycis (L.)

Mediterranean and eastern Atlantic between Biscay and Cape Verde.

Similar to *Phycis blennoides* but there are no elongate rays in the dorsal fin and the pelvic fins reach back no further than the origin of the 1st anal fin. 1D 8-9; 2D 60-63; A 58-64.
Blackish or reddish brown above, lighter on sides and belly.
Up to 25 cm., rarely 60 cm.

Lives from moderate depths to 600 m. Not very common although locally they may be more abundant than *Phycis blennoides*.

Hake p. 173
Merluccius merluccius (L.)

North Atlantic from Lofotens southwards to N. W. Africa, British Isles (especially the west coasts), Mediterranean, Black Sea.

Body elongate; *jaws*, lower longer than upper; *teeth* prominent; *lateral line* almost straight; *barbels* absent; *fins* 2 dorsal, 1 anal. The second dorsal fin origin above the origin of the anal fin; both these fins are long and of uniform height. The pectoral fins are long reaching back to the origin of the anal fin. 1D 9-10; 2D 37-40; A 36-40.
Blackish-grey on back, silvery grey on flanks. Median fins have dark margins.
Up to 100 cm. in northern waters, up to 35 cm. in the Mediterranean.

A species of considerable commercial importance but generally lives too deep to be seen by divers. Lives in depths from 100-1000 m. although the young may be found in shallower water. They feed almost entirely on pelagic fish such as mackerel, herring and sardine implying that they spend much of their life above the bottom.
Breed in summer and autumn in the North Sea, earlier in the Mediterranean. There are well defined spawning grounds but at very variable depths. The eggs are pelagic.

Ling p. 173 **pl. 17**
Molva molva (L.)

North-east Atlantic south to Biscay.

Body elongate; *jaws*, upper longer than lower; *barbel*, conspicuous barbel on chin and one small barbel near each nostril; *fins*, 2 dorsal fins and 1 anal. Tip of pelvic fin does not extend beyond tip of pectoral. 1D 14-15; 2D 62-65; A 58-61.
Mottled brown or green with a bronze sheen. There is a dark blotch at the rear end of both the 1st and 2nd dorsal fins. Both 2nd dorsal and anal fins are outlined in white. The young have a yellowish-olive ground colour broken up with lilac iridescent lines.
Up to 100 cm., rarely up to 220 cm.

A fairly important commercial fish, it chiefly lives in water of 300-400 m., but immature fish are quite often seen as shallow as 10 m. where they tend to favour crevices between rocks. These immature fish live in shallow water for some 2 years (when they may reach 50 cm.) and then move into deeper water.
The Ling's diet is not known in detail but it feeds extensively on small fishes together with some crustaceans and echinoderms.

Mediterranean Ling p. 173
Molva elongata (Otto)
= *Molva macrophthalma* (Rafinesque Smaltz)

Mediterranean and Atlantic north to S. W Ireland.

Differs from *Molva molva* in the following: *body* longer and more eel-like; *jaws*, lower longer than the upper; *barbel* forked; *fins*, tip of pelvic reaches past tip of the pectoral; *eye* small; vertebrae 82-84.
Grey-brown above, silvery below. Dorsal, anal and tail fins have violet edges. The pelvic fins are blue.
Up to 90 cm.

On sandy mud between 200 and 1000 m.

Tadpole Fish   p. 173
Raniceps raninus (L.)

North Sea and all coasts of England and Ireland.

Body tadpole-shaped. The breadth of the head nearly equals its length and occupies 1/3 of the total length of the body; *barbel,* there is a single short barbel on the lower jaw; *fins,* 1st dorsal has only 3 rays, 2nd dorsal and anal are long, and of uniform height. 2nd ray of the pelvic fin is elongate, extending back at least as far as the vent. 1D 3; 2D 63-64; A 57-60.
Dark brown with lilac highlights and whitish beneath. The dorsal, tail and anal fins are dark with a light fringe. The lips are light in colour and are probably very conspicuous underwater.
Up to 30 cm.

Solitary fish which live from immediately beneath the low-tide mark to about 100 m. They feed on bottom living animals such as molluscs, crustacea, annelids and echinoderms. The smell of the fresh fish is very unpleasant. They spawn in summer in 50-75 m. at temperatures of 10-12°C.

Shore Rockling   p. 174 **pl. 18**
Gaidropsarus mediterraneus (L.)
= *Motella tricirratus*
= *Onos tricirratus*

Mediterranean. Atlantic coasts of Europe, west and south coasts of England, south Ireland.

Body elongate; *mouth* small, extending to rear edge of eye; *barbels* 3; *fins,* 1st dorsal is composed of very fine rays set in a groove on the back; 2nd dorsal and anal fins are long and of uniform height extending back to the tail fin. Pectoral fin has 17 rays. 1D I + ∞; 2D 53-63; A 43-53.
Colour varies with habitat, usually brown or dark brown. Never mottled except occasionally on the lower part of head. Young silver with bluish or greenish back.

Up to 50 cm.

Lives amongst rocks, often in tide pools and not deeper than 30 m. They feed upon small bottom living fishes, molluscs, crabs etc. Spawn close inshore in summer. The young (Mackerel Midges) live just beneath the surface and by the autumn reach about 4 cm. in length. They then settle on the bottom in shallow water.

Three-Bearded Rockling   p. 174
Gaidropsarus vulgaris (Cloquet)   **pl. 19**

Mediterranean, North East Atlantic north to Faroes.

Differs from the shore rockling (*Gaidropsarus mediterraneus*) in the following characters: *Mouth* large extends past the eye; *fins,* 20-22 rays in pectoral fin. 1D I + ∞; 2D 56-64; A 46-52.
Very characteristic leopard-like pattern of bold dark spots upon a brick red ground. Fish shorter than 10 cm. is not spotted.
Up to 55 cm.

The smaller specimens always occur amongst rocks although larger specimens may be found on soft bottoms nearby. They live from just beneath low-tide level down to about 120 m.
They spawn in winter, the eggs float and the young swim in the surface waters. During this so-called 'Mackerel Midge' stage they have a green back and silvery flanks and are often found swarming in great numbers. Once they reach about 3 cm. the young take up a bottom living existence. Here they feed upon bottom living molluscs, crustacea and fishes.

Four-Bearded Rockling   p. 174
Rhinonemus cimbrius (L.)

North Sea and Baltic, North Atlantic south to Biscay.

Body elongate; *barbels* four, 2 on each of the front nostrils, 2 on the chin; 1 in the

180

middle of the upper jaw; *fins*, 1st dorsal ray long almost equalling the head length. 1D I+∞; 2D 45-50; A 37-41.

The colour is a uniform brown but is sometimes chestnut brown, sometimes very dark. Pectoral and dorsal fins are light grey. The 2nd dorsal and anal fins are grey with a lighter fringe. There is a dark oval blotch at the hind end of both the 2nd dorsal and the anal fins.

Up to 40 cm.

A deep water fish, rarely more shallow than 50 m.

Five-Bearded Rockling   p. 174
Ciliata mustela (L.)

North-east Atlantic south to Portugal

Body elongate; *mouth* rather small reaches back to the hind edge of the orbit; *barbels* five, one on each of the front nostrils, 2 on the upper lip and 1 on the lower lip. There are no small fringing barbels on the upper lip; *fins*, 1D I+∞; 2D 50-55; A 40-41.

Colour uniform, ranging from reddish-brown to dark brown.

May reach 45 cm. in length but usually no longer than 25 cm.

Live mostly between the tides although the larger fish may live beneath the low-tide mark. They tend to prefer soft bottoms near rocks. They spawn offshore in the winter. The surface living 'Mackerel Midge' stage lasts until early summer when the silvery colour is lost to the drab brown of the bottom living form. They feed on small fish, crustacea and molluscs.

Northern Rockling   p. 174
Ciliata septentrionalis (Collett)

North Sea and north-east Atlantic south to Cornwall.

Resembles the Five-Bearded Rockling except that the mouth is large, extending well behind the orbit and the upper jaw is fringed with supplementary barbels.

Apparently rare it is caught from 10-100 m. It is possible that this fish has been confused with the Five-Bearded Rockling.

Order: SOLENICTHYES

Usually soft-rayed fishes with the mouth at the end of a long tubular snout.

Family:

MACROHAMPHOSIDAE

Unusual amongst tube-mouthed fishes in having a deep body and a long spine in its dorsal fin.

Snipe Fish; Trumpet Fish
Macrohamphus scolopax

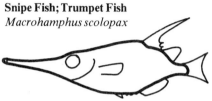

Very long snout;
red in colour;
2nd ray of 1st dorsal fin elongate.
15 cm.

Eastern Atlantic north to Norway, Mediterranean.

Body deep and flattened sideways; *snout* elongated to form a tube; *mouth* very small and situated at the end of the snout; *teeth* none; *fins*, 2 dorsal fins, the 1st consists of spiny rays and is situated in the rear half of the body. The 2nd ray is greatly elongated and has a toothed rear edge. 1D VI-VIII; 2D 11-13; A 18-20.

Back red, sometimes olive, sides pink, belly pinkish-silver.
Up to 15 cm.

Mid-water fish found between 20-200 m.

but also down to 600 m. Not common except in Biscay and the Mediterranean where they are found in shoals. Usually found over sand and mud bottoms; They feed on small planktonic animals.

Family:

SYNGNATHIDAE

The **Pipefishes** and **Sea Horses** have an external skeleton made up of bony plates. The snout is extended into a long narrow pipe and the mouth which is very small is situated at the tip. All fins except the dorsal fin are very much reduced or absent. The Sea Horses differ from the Pipe Fishes in having prehensile tails and the head set at an angle to the body.
The eggs and larvae are carried by the male until they hatch. In members of the genus *Entelurus* the eggs are attached along the outside of the belly. In *Nerophis* the eggs are held in a narrow groove whilst males of the genus *Syngnathus* have a brood pouch running along the belly which opens to the outside world through a longitudinal slit. The brood pouch of Sea Horses is almost entirely enclosed by skin and there is only a small opening.

Sea Horse *Hippocampus ramulosus*
p. 184

Sea Horse *Hippocampus hippocampus*
p. 184

Snout more than 1/3 head length.
15 cm.

Snout 1/3 or less head length.
15 cm.

Spotted Pipefish *Nerophis maculatus* p. 187

No pectoral or tail fins;
regular spotted pattern,
no longitudinal crest on snout;
dorsal fin extends over 8-9 segments.
30 cm.

p. 186
Straight-Nosed Pipefish *Nerophis ophidion*

No pectoral or tail fins;
snout has longitudinal crest giving a straight profile;
dorsal fin extends over 10-12 segments.
30 cm.

p. 186
Worm Pipefish *Nerophis lumbriciformis*

No pectoral or tail fins;
snout short and up-tilted;
back very dark.
17 cm.

Snake Pipefish *Entelurus aequorius*   p. 186

Pectoral fin absent;
tail fin small;
dorsal fin set forward.
60 cm.

p. 186
Dark-Flank Pipefish *Syngnathus taenionotus*

Outline of head from snout to nape flat;
dark line on flank.
20 cm.

Syngnathus phlegon   p. 185

Snout long and tubular (more than 1/2 total head length);
back rough.
20 cm.

Syngnathus pelagicus   p. 185

Snout long and tubular (more than 1/2 total head length);
tail fin pointed.
17 cm.

p. 185
Short-Snouted Pipefish *Syngnathus abaster*

Snout with longitudinal crest;
skin smooth.
17 cm.

p. 185
Nilsson's Pipefish *Syngnathus rostellatus*

Snout short, less than 1/2 head length;
13-17 rings in front of vent.
17 cm.

p. 184
Deep-Snouted Pipefish *Syngnathus typhle*

Snout laterally compressed.
35 cm.

Great Pipefish *Syngnathus acus*   p. 184

Hump on nape;
snout tubular, more than 1/2 head length;
17-22 rings in front of vent.
30 cm.

Sea Horse   p. 182 **pl. 167**
Hippocampus ramulosus Leach
= *Hippocampus guttulatus guttulatus* Cuvier

Mediterranean, western Atlantic from the English Channel to west Africa. Rarely North Sea.

Body bent at the neck, 48-50 body segments often with bony thickenings on the ridges and fleshy tentacles along the back; *tail* prehensile; *snout* long, more than 1/3 head length; *fins* dorsal and pectoral and anal fins, no tail fin. D 18-21; A 4; pectoral fin rays 15-16.
Greenish or brownish, sometimes with white spots on the corners of the body segments.
Up to 15 cm.

Lives in shallow water amongst sea weeds, in particular amongst sea grass (*Posidonia*) meadows where it clings to the vegetation with its tail. Also found in brackish water. Sea Horses are very rarely seen by divers even where they are quite common, presumably because their irregular outline makes them very difficult to see. They swim in an upright position using their dorsal fin as their sole means of propulsion. Spawn in late spring and summer.

Sea Horse   p. 182
Hippocampus hippocampus (L.)

Mediterranean, French and Spanish coasts, probably not as far north as Britain.

Resembles *H. ramulosus*, the only certain point of difference is the snout which is short, 1/3 the head length or less; 45-47 body rings; fins D 16-17; A 4; pectoral fin rays 13-15.

This species appears to favour rather muddy bottoms. Spawning is in spring and summer. The male follows the female about presenting his brood pouch to her. The transfer of eggs normally begins at night, the part-ners swim side by side and the female transfers a few eggs at a time. The young fish are expelled from the brood pouch fully developed after 4-5 weeks.

Great Pipefish   p. 183 **pl. 20**
Syngnathus acus L.   **pl. 165**
= *Syngnathus tenuirostris* Rathke

North-east Atlantic especially west coast of British Isles and Ireland, southern North Sea, English Channel, Biscay, Mediterranean.

Snout long, more than 1/2 total head length but maximum height less than eight times the maximum length. No ridge along the top margin of the snout; *head*, the head behind the eye has a conspicuous bump; *body rings* 17-22 between the base of the pectoral fin and the vent, polygonal in cross section; *fins*, tail and pectoral fins present. The dorsal fin extends over 7-9 body rings.
Dark or light grey brown or greenish above, lighter below. There are darker mottlings usually in the form of cross-bands.
Up to 30 cm., sometimes 50 cm.
Lives on most kinds of bottom often amongst sea grass and sea weed, usually more shallow than 15 m. although it may go deeper. They are able to live in brackish water and are often found in river estuaries. Breeding occurs in summer. The young fish leave the brood pouch after about five weeks.

Deep-Snouted Pipefish; Broad-Snouted Pipefish   p. 183 **pl. 166**
Syngnathus typhle (L.)

Mediterranean, north-east Atlantic north to Norway, Baltic.

Snout deep and laterally compressed, usually equalling the head in depth throughout its length; *profile*, the top of the head and snout form a continuous line; *body rings*, 16-20 (usually 19) rings in front of the vent; *fins* pectoral and tail fins present. The dorsal fin extends over 7-9 body rings and has 29-

39 rays.

Light brown or greenish brown above, lighter below, sometimes with sparse vertical stripes. Head and snout with darker spots.

Up to 35 cm.

Very tolerant of brackish water it is common in the Baltic, the Dutch Waddenzee and the Black Sea. It assumes an upright stance typically amongst sea grass but also amongst seaweeds.

Breeds in summer, the young fish are expelled from the brood pouch after about 4 weeks. Feeds on crustacea and small fishes.

Nilsson's Pipefish; Lesser Pipefish   p. 183
Syngnathus rostellatus Nilsson

North Sea, English Channel, west coast of England and Ireland, Biscay.

Resembles *Syngnathus acus* except that the snout is short (less than 1/2 total head length); *profile*, the nape is broadly convex but does not amount to a hump-like crest as in *Syngnathus acus*; *body rings* 13-17 rings in front of the vent.

Up to 17 cm.

Found over sandy bottoms down to about 18 m. especially amongst floating or attached seaweed, sometimes it is seen swimming freely. It can live and breed in brackish water.

Short-Snouted Pipefish   p. 183
Syngnathus abaster Risso
= *Syngnathus agassizi*

Mediterranean and west coasts of Spain and Portugal.

Snout short and tubular but with a pronounced high membrane-like crest running along its upper margin; *body rings* 14-18 (usually 16) rings in front of the vent; *skin* smooth to the touch when rubbed from tail to head; *fins* tail and pectoral fins present. The dorsal fin extends over 7-11 segments

and has 23-34 (usually about 28) rays.

Greyish or greenish brown above, paler below. There may be a row of dark or light spots along the flanks which sometimes show darker cross-bands.

Up to 17 cm.

On sandy bottoms, often near estuaries.

Syngnathus pelagicus L.   p. 183
= *Syngnathus agassizi* Canestrini

Atlantic, Indian and Pacific oceans, European Atlantic coasts, rarely in Mediterranean.

Snout tubular, more than 1/2 total length of head; *body rings* are smooth, 13-18 in front of the vent, 33-42 behind; *fins*, pectoral fin present. The tail fin comes to a point. The dorsal fin has 30-45 rays (typically 40) and extends over 9-11 body rings.

Body a uniform greyish brown sometimes speckled or with indistinct crossbands. There is a dark longitudinal stripe in front of each eye.

Up to 17 cm.

Usually found in association with floating masses of seaweed in the open oceans.

Syngnathus phlegon Risso   p. 183

European Atlantic coasts and Mediterranean.

Snout tubular and elongate, longer than 1/2 the total head length; *body rings* 17-19 in front of the vent, 47-50 behind. The ridges along the body are markedly abrasive to the touch when the fish is stroked from the tail to the head; *fins* tail and pectoral fins present. The dorsal fin extends over 12-14 body rings (2-3 in front of the vent) and has 38-46 rays.

Either bluish with a row of dark spots along the flanks or silvery yellow with darker vertical stripes.

Up to 20 cm.

An open water species probably found from the surface to 600 m. or more.

Dark-Flank Pipefish   p. 183
Syngnathus taenionotus Canestrini

Snout long and tubular; *profile*, there is no raised hump on the nape and the line of the head from above the pectoral fin to the snout tip is flat; *fins*, the dorsal fin extends over 7-11 body rings. Pectoral and tail fins present.
Grey often with a dark longitudinal stripe extending along the upper flank from the pectoral fin to just behind the vent.
Up to 20 cm.

Found amongst seaweed in sheltered areas.

Snake Pipefish   p. 183
Entelurus aequorius (L.)   **pl. 163**
= *Nerophis aequorius* (L.)
= *Syngnathus aequorius* L.

Eastern Atlantic from Scandanavia to the Azores; rare in the Mediterranean.

Body long and thin and rounded rather than polygonal in cross-section; *body rings*, 28-31 in front of the vent and 60-70 behind; *fins*, pectoral fin absent, the anal fin is very much reduced with only 4-6 rays. The dorsal fin set forward covering 8 body rings in front of the vent and 3 behind. It has 37-44 rays.
Yellowish or brownish grey. On the flanks there are numerous silvery transverse markings often with dark borders. There is a reddish longitudinal band running from the tip of the snout through the eye to the gill cover.
Males reach 40 cm., females 60 cm.

Found amongst marine plants in shallow water down to about 30 m. In summer they are sometimes found swimming near the surface far from the coasts, particularly near floating seaweed. They breed in summer, the young are not fully developed when they leave the brood pouch and live a free swimming existence in open water.

Worm Pipefish   p. 183
Nerophis lumbriciformis (Jenyns)   **pl. 164**

North-east Atlantic from Scotland to Morocco, English Channel, North Sea.

Body very long and slender and rounded in cross-section; *snout* short and markedly tilted upwards; *body rings* indistinct, 24-28 in front of the vent, 17-18 behind; *fins*, there are no pectoral or tail fins. The dorsal fin is set rather far back extending over 2-3 body rings in front of the vent and 4-6 behind.
Very dark greenish grey or dark brown above mottled white below.
Up to 17 cm.

The Worm Pipefish is unusual in living in very shallow water, even between tide marks, where they are usually found under stones. They prefer rocky areas where there is a luxuriant algal cover.

Straight-Nosed Pipefish   p. 183
Nerophis ophidion (L.)

North-east Atlantic and Mediterranean.

Body long and slender, rounded rather than polygonal in cross-section; *snout* is half the total head length with a raised longitudinal crest running from between the eyes to the tip of the snout; *body rings* there are more than 25 body rings on the trunk in front of the vent; *fins*, no tail or pectoral fins. The dorsal fin extends over about 11 body rings (3-5 in front of the vent, 6-8 behind), it has 33-38 rays.
Colour very variable, dark green, olive, brownish or blackish above with dark vertical lines, pale whitish green, yellow or blue below. Sometimes the front half of the fish has lighter spots arranged in regular vertical rows.
Up to 30 cm.

Lives close to shore amongst seaweeds, especially sea grass and thong weeds (*Chorda* and *Himanthalia*).
Breeding occurs in summer, the eggs are

held in a groove along the belly of the male. The young are released in a relatively undeveloped state and live a free-swimming life for 3-4 months.

Spotted Pipefish   p. 183
Nerophis maculatus Rafinesque

Mediterranean and eastern Atlantic from Portugal to the Azores.

Body long and slender, rounded rather than polygonal in cross-section; *snout* rounded with no longitudinal crest; *body rings*, there are 21-23 body rings on the trunk in front of the vent and 65-74 behind; *fins* no pectoral or tail fin. The dorsal fin extends over 8-9 segments (2-4 in front of the vent, 4-5 behind). It has 24-30 rays.

Greyish, brownish, greenish or greyish-red with rows of yellowish spots with darker edges arranged in vertical bars, sometimes coalescing into vertical bands.

Males up to 20 cm., females up to 30 cm.

Order: ZEIFORMES

Deep laterally flattened bodies with tall dorsal fins, large eyes and protrusible mouths.

Family:

ZEIDAE

John Dory   pl. 21, 168
Zeus faber (L.)

Unmistakeable shape;
round spot on side of body.
40 cm.

Mediterranean and Eastern Atlantic from Scotland to West Africa. Rare in the North Sea.

Body round and very flattened laterally; *mouth* protractile; *scales*, on each side of the base of the dorsal and anal fins there is a row of armoured scales with sharp spines; *fins*, there is a single dorsal fin. The spiny rays are long and the tissue of the fin membrane is prolonged into a long filament. The anal fin has 4 strong spiny rays. D IX-X, 21-25; A III-IV, 20-23.

Grey or yellowish with indistinct mottling. There is a dark central spot with a lighter margin.

Up to 40 cm., rarely up to 50 cm.

A solitary fish which ranges in depth from a few metres to 200 m. It is often trawled up from sandy ground but in the summer the young fish are sometimes seen by divers amongst the weed growing on rocky bottoms. The John Dory feeds almost entirely on fishes which it slowly approaches (it is not a strong swimmer), and then it accelerate with a jerky motion, swinging its jaws forward to catch its prey.

They spawn in summer in water shallower than 100 m. but not further north than the Irish sea or further east than the western

English Channel. In the warmer waters of its range they spawn in spring. The older fishes longer than 20 cm. remain near their spawning grounds, but the younger fishes travel further afield.

Very good to eat although its grotesque appearance may be disconcerting. This is one of the fish that is supposed to taste better after it has been dead for two or three days.

Family:

CAPROIDAE

Boar Fish   pl. 22
Capros aper (L.)

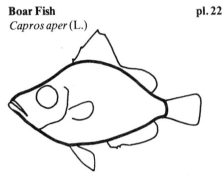

Body deep oval; *eye* large; *profile* concave with a long snout; *mouth* can be protruded; *fins*, dorsal fin in two parts. The dorsal spines are long and robust. There is one pelvic spine and 3 anal spines. 1D IX; 2D 23; A III, 23.
Yellowish ochre or red occasionally with vertical yellowish bars.
Up to 16 cm.

Red or orange body; long snout.
16 cm.

Mediterranean and Eastern Atlantic north to Ireland.

A deep water fish found from 100-400 m. Caught over both soft and rocky bottoms. Occasionally it is found in shallower water but these individuals may have been carried up on upwelling currents and are not part of a permanent population.

Order: PERCIFORMES

A very large order of Perch-like fishes. 1 or 2 dorsal fins. The first dorsal fin is usually spined or if there is only one dorsal fin the front part is spined.

Sub-order: MUGILOIDEI

There are two dorsal fins always well separated. The pelvic fins are on the abdomen and have one spiny ray and five soft rays.

Family:

SPHYRAENIDAE

Long cylindrical body with two well separated dorsal fins. The mouth is large and contains numerous large teeth. The lower jaw extends beyond the upper.

Barracuda   **pl. 169**

Sphyraena sphyraena (L.)
= *Esox sphyraena* L.

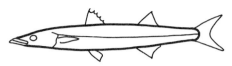

Long slender body;
2 dorsal fins;
powerful jaws;
vertical stripes in adult.
50 cm.

Mediterranean; East Atlantic north to Biscay.

Body long and slender; *snout* long; *mouth* long, reaching back to the front edge of the eye; the lower jaw is longer than the upper

and has a small lobe at the end. When the mouth is shut the lobe is arranged in such a way that it produces a smooth, streamlined outline; *teeth* many and pointed; *fins*, 2 dorsal fins well separated. Anal fin similar to and opposite 2nd dorsal. Tail fin forked. Pectoral fin short but slender. 1D V; 2D I, 9; A I, 9.

Back greenish or brownish. Adult fish have about 24 darker vertical stripes. Lower flanks and belly silver.
Up to 50 cm.

These fish are extremely voracious and aggressive. They are found in shoals sometimes just underneath the surface but also much deeper, especially above sand. They are carnivorous and feed on fish which they hunt and attack as a group. Breeding occurs at the end of spring and the beginning of summer.

family:

ATHERINIDAE

The Sand Smelts are schooling species of shallow water. They are frequently mistaken for sardines from which they differ in having two dorsal fins.
The body is covered with scales, the mouth is terminal and is directed steeply upwards. The first dorsal fin has 6-9 spiny rays. The species are difficult to distinguish.

Sand Smelt   **pl. 170**

Atherina presbyter Valenciennes

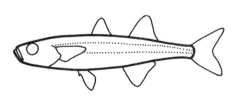

2 well spaced dorsal fins;
bright silver stripe along flanks.
15 cm.

East Atlantic; English Channel; North Sea; western Mediterranean.

Body rather long and slender; *head* length fits 41/2 times into total body length; *mouth* rather large and extends to about the front edge of the eye; *skin* covered with small scales, about 53-57 scales from the top of the pectoral fin to the tail base; *fins*, 2 dorsal fins, the 2nd rather larger than the 1st opposite to and slightly shorter than the anal fin. Tail fin forked.

The back is greenish, each scale has small black dots along the rear edge. The flanks have a bright silver stripe, only a little less wide than the eye. Belly silver.
Up to 15 cm.

These are common fish found near coasts and often in estuaries and areas of low salinity. Usually in shoals. They spawn in spring and summer often in fresh or brackish water and the eggs are attached to seaweeds, stones or sand by long filaments. They feed on small crustacea, fish larvae and fry. They are important fish as they form part of the food for commercially important species like herring.

Caspian Sand Smelt
Atherina mochon Valenciennes

Mediterranean; Black Sea; rarely west and south England.

Similar to *Atherina presbyter* but may be distinguished by the distance between the dorsal fins which equals twice the eye diameter, the snout is fairly pointed and there are less than 50 scales along the lateral line. Up to 12 cm.

The habits of this fish are similar to *Atherina presbyter*. They prefer fresh or brackish water for breeding. Spawning occurs from May throughout the summer. The eggs are attached to seaweeds or rocks.

Sand Smelt
Atherina hepsetus L.

Mediterranean; Black Sea.

Similar to *Atherina presbyter* but may be distinguished by the distance between the 2 dorsal fins which is more than twice eye diameter, a rather pointed snout and about 60 scales along the lateral line. Up to 15 cm.

The habits of this fish are similar to *Atherina presbyter*. Breeding occurs from January to August and this species appears to prefer small sheltered bays where they spawn from 4-8 m. The eggs are attached to rocks and weeds.

Boyer's Smelt
Atherina boyeri Risso

Mediterranean.

Similar to *Atherina presbyter* but may be distinguished by the distance between the dorsal fins which is about equal to 1 eye diameter, there are 55-58 scales along the lateral line and the snout is blunt. Up to 10 cm.

The habits of this fish are similar to *Atherina presbyter*. Breeding occurs from March to August.

Family:

MUGILIDAE

Like the Red Mullet (to which they are not closely related) the **Grey Mullet** are an important food fish making up in quantity what they lack in quality. They tend to live near the coasts where they congregate in large schools feeding on the algae growing over rocks, harbour walls, sewage outfalls etc. The smaller fish are often seen swimming just beneath the surface of the water. The Grey Mullet are very tolerant of low salinity and are common in river estuaries.

The body is long and slender with large scales. The mouth is small and terminal with small teeth. There are two widely spaced dorsal fins of which the first has 4 spiny rays. A very homogeneous family; the species can be difficult to distinguish.

190

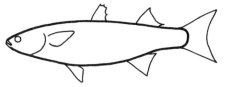

Transparent eyelid
Common Grey Mullet   *Mugil cephalus*
No transparent eyelid—*see* **a**

a Upper lip swollen, width more than 1/2 eye diameter—*see* **b**
Upper lip width less than 1/2 eye diameter—*see* **c**

b 1st dorsal fin more or less equal to anal
Thick-Lipped Grey Mullet   *Mugil labrosus*   p. 191
1st dorsal fin about 2/3 length of anal fin
Mubil labeo   p. 192
c Scales on top of snout not extending to upper lip
Golden Grey Mullet   *Mugil auratus*   p. 192
Scales on top of snout extending to upper lip—*see* **d**

d Pectoral fin rounded; grey often with black spot at base
Thin-Lipped Grey Mullet   *Mugil capito*   p. 192
Pectoral fin not rounded and yellowish, no black spot at base
Leaping Grey Mullet   *Mugil saliens*   p. 192

Common Grey Mullet
Mugil cephalus L.

Distribution world-wide but it is rare in the North Atlantic north of Biscay.

The only Grey Mullet with transparent (adipose) eyelids.
Up to 50 cm., rarely 65 cm.

Found around rocky and sandy coasts especially near freshwater outlets. They swim in dense shoals, a habit particularly marked in the young. The males are only 2/3 the size of the females and in the breeding season the males have a distinct pearly sheen. It is not certain where they spawn but mature fish are caught throughout the world in summer and autumn.

Thick-Lipped Grey Mullet   **pl. 172**
Mugil labrosus

Mediterranean and East Atlantic south to Senegal and north to Norway.

Lips, the upper is swollen and has rows of small blister-like protuberances in older specimens; *fins,* the 1st dorsal is about equal in length to the anal.
Up to 60 cm.

The most abundant Grey Mullet of north Europe where they are common inshore and in brackish water. They are most abundant in spring and summer where they feed on the diatoms and other algae coating the rock surface. Their diet also includes animals such as molluscs that are attached to the rock surface. In the winter feeding stops and they appear to enter a state of hibernation. It is uncertain whether they ever feed further north than Biscay.

In the Mediterranean they spawn from December to March and in Biscay from January to April.

Mugil labeo Cuvier   p. 191

Mediterranean and East Atlantic between Gibraltar and Senegal.

Lip, upper lip thick without protruberances; *head* very blunt; *fins,* dorsal only 2/3 length of anal.
The upper part of the flanks have longitudinal yellow stripes.
Up to 30 cm.

Not a common species and its habits are not well known. The adults spawn in the sea and the fry are found amongst rocks where there is a strong surge.

Golden Grey Mullet   p. 191
Mugil auratus Risso

Mediterranean, East Atlantic south to Senegal and rarely as far north as North Europe.

Lips less than 1/2 eye diameter; *fins,* 1st Dorsal about equal in length to anal. The pectorals are long and not rounded.
On the cheeks and gill cover there is a well-marked golden blotch.
Up to 45 cm.

Rare in northern waters. In the Mediterranean they have some commercial importance. They spawn from July to October.

Thin-lipped Grey Mullet   p. 191
Mugil capito Cuvier

Throughout entire Mediterranean and East Atlantic but are uncommon in North Europe.

Lips, upper lip less than 1/3 eye diameter; *fins,* pectoral fin is short and rounded.
Grey pectoral fin, often with a black spot at its base.
Up to 60 cm.

Chiefly a brackish water species. They live in shallow water although in the northern part of their range they are believed to migrate offshore in winter.

Leaping Grey Mullet   p. 191
Mugil saliens Risso

Mediterranean and East Atlantic from South Africa to Portugal.

Body slightly more slim than the other Grey Mullets; *jaws,* hind edge of the jaws are visible when the mouth is shut not being hidden by the bone of the front gill cover; *scales* on the top of the head extend to the lip; *fins,* the pectoral fin is not rounded and has a yellow tinge.
Up to 30 cm.

Sub-order: PERCOIDEI

There are spiny rays in the fins and the pelvics are well developed. The scales are mostly of the ctenoid type, i.e. the hind edge of each scale is minutely toothed.

Family:

APOGONIDAE

Small fishes with two short well-separated dorsal fins. Two spiny rays in the anal fin. The mouth and eye are generally large.

Cardinal Fish   **pl. 24**
Apogon imberbis (L.)
= *Apogon ruber* Lacépède

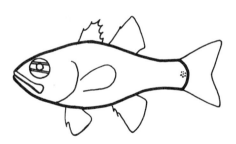

Red;
fins not elongate.
15 cm.

Mediterranean; East Atlantic around Madeira, Azores and the Canaries.

Body short and deep; *head* large; *eyes* very large; *front gill cover* toothed; *mouth* large; *teeth* small and pointed, similar in both jaws; *fins*, 2 dorsal fins the 1st smaller than the 2nd which is opposite and similar to the anal fin. 1D VI; 2D I, 9; A II, 8.

Bright orange-red with small black dots irregularly scattered over the surface. There is a darker area on the tail stalk which is sometimes present as 2 or 3 spots. The tips of the dorsal and anal fins are darker. Under surface lighter. The eye has two light horizontal bands.

10-15 cm. in length.

These fish are found from 10-200 m. They are commonly seen in small groups in caves or crevices in rock falls where they hover near the entrances. At night they venture further afield. Breeding occurs in the summer, with mating being accompanied by quivering movements. After spawning the eggs are bound together into one mass with threads which are present at one end of each egg. The male fish then takes the egg mass into his mouth where it remains until just before the eggs are hatched when he spits them out. Feeds on small crustacea, larvae and eggs.

Family:

SERRANIDAE

This family includes the well known **Groupers** and **Combers**. They are chiefly found close to the coasts and are carnivorous. Many species may change from one sex to the other as they get older and at least one species, the Comber (*Serranus cabrilla*) is able to fertilize its own eggs and is thus a true hermaphrodite.
The gill cover has three spines. There is generally a single dorsal fin but in a few species it is divided. The anal fin has 3 spines and 7 soft rays. One spine and five soft rays in the pelvic fin.

Bass   p. 194
Dicentrarchus labrax (L.)   **pl. 171**
= *Morone labrax* (L.)

Mediterranean; East Atlantic north to Norway south to Senegal; Baltic; southern North Sea; English Channel.

Body long and oval; *head* fairly long, 5-6 times the diameter of the eye; *front gill cover* has a toothed rear edge; *gill cover* with 2 spines; *scales* small, 65-80 along the lateral line; *fins* there are 2 dorsal fins the 1st dorsal is entirely separate from the 2nd. Tail fin slightly forked. 1D VIII-IX; 2D I, 12-13; A III, 10-11.

A silvery fish, grey on the back, lighter on the flanks and white on the belly. Fish up

Bass *Dicentrarchus labrax*   p. 193

Dorsal fins separate;
no spots.
100 cm.

Spotted Bass *Dicentrarchus punctatus*
p. 195

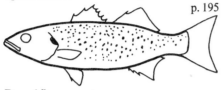

Dorsal fins separate;
dark spots on body;
black spot on gill cover.
60 cm.

Wreck Fish *Polyprion americanum*   p. 195

Heavy body, deep head;
scales extend onto dorsal, anal and tail fins;
bony ridge on gill cover.
2 m.

Grouper *Epinephelus guaza*   p. 196

Round tail;
orange rim to dorsal fin.
1.4 m.

Grouper *Epinephelus alexandrinus*   p. 196

Square tail;
very protruding lower jaw;
4-5 dark longitudinal stripes in young.
80 cm.

Comber *Serranus cabrilla*   p. 196

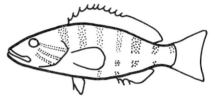

7-9 dark vertical bands crossing
2-3 longitudinal pale blue stripes.
20 cm.

Painted Comber *Serranus scriba*   p. 197

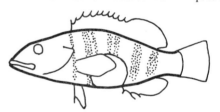

Large blue spot on flank.
25 cm.

Brown Comber *Serranus hepatus*   p. 197

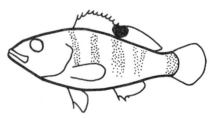

Black spot on dorsal fin.
13 cm.

to 10 cm. may have black spots on the sides and back.
Up to 100 cm.

Found at all depths from near the surface to below 100 m. The adults are usually solitary but the young fish do form schools. During the summer they migrate to inshore water where they are frequently found in estuaries and may penetrate far up rivers. They are found over bottoms ranging from rocks and sand to shingle and mud. They appear to occupy the same localities for many months and observations by divers indicate that they may occupy well defined feeding territories even between the tides. They are carnivorous, the young fish feeding upon small fish and crustacea and the adults to a greater extent upon fish although they may also eat worms, cephalopods and crustacea. In the Mediterranean they spawn in January to March and in the Channel from March to June. They are edible and a favourite angler's fish.

Spotted Bass p. 194
Dicentrarchus punctatus (Bloch)

Mediterranean; East Atlantic between Biscay and Senegal, common around Gibraltar.
Body long and oval similar to the Bass (*Dicentrarchus labrax*) but with a slightly shorter head and snout; *snout* about 4 1/2 times the diameter of the eye; *teeth*, the vomer teeth are T shaped; *front gill cover* has a toothed rear edge; *scales* 57-68 along the lateral line; *fins*, 2 dorsal fins the 1st fin entirely separate from the second. Tail fin slightly forked. 1D VIII-IX; 2D I, 11-12; A III, 10-12.
Silvery but with a dark blue back. Adults with black spots scattered irregularly over the back and sides. There is a conspicuous black spot on the gill cover.
Up to 60 cm.

The habits of this fish are very similar to those of *Dicentrarchus labrax*. They may be found over sand or mud with sand and rocks and seem to prefer brackish water. They are carnivorous feeding upon fish, molluscs and crustacea.

Wreck Fish p. 194 pl. 173
Polyprion americanum Bloch & Schneider

Mediterranean; throughout Atlantic but only occasionally north of Ireland; English Channel.

Body deep and heavily built; *head* large; *profile* fairly steep with a slight concavity above the eyes with one group of lumps between the eyes and another on the forehead; *jaws*, the lower jaw is clearly longer than the upper; *teeth* are small, strong, sharp and conical; *front gill cover* with a toothed rear edge; *gill cover* has a distinct bone running lengthways across it; *scales* are small and cover the head and body and extend onto the base of all the fins but to a greater extent onto the dorsal, anal and tail fins; *fins* 1 dorsal fin with very strong spines; tail fin rounded or straight. D XI, 11-12; A III, 8-10.
The adult fish has a dark brown back, lighter sides and yellowish belly. Sometimes with faint lighter blotches on the back and sides. Young fish may have scattered irregular large white and smaller black blotches. The edge of the tail fin is white.
Up to 2 m.

These fish live in deep open water but are often encountered around floating wreckage or seaweed. They usually live between 100-200 m. but occasionally as deep as 1000 m. They may either be solitary or in large numbers particularly when encountered in association with floating debris. Usually these schools of fish are young specimens, the adults lead a more nomadic, bottom living existence. Their habit of following floating material does sometimes bring the fish close inshore. However, in the Mediterranean there is a general movement inshore during the summer and autumn. They spawn in the Mediterranean between January and April and feed on fish, crustacea and molluscs.

Grouper; Dusky Perch   p. 194 **pl. 26,**
Epinephelus guaza (L.)   **80**
= *Serranus gigas*

Mediterranean; occasionally East Atlantic, rare north of Biscay.

Body oval; *jaws*, the lower slightly in front of upper; *teeth* both jaws contain long, strong sharp and moveable teeth; *front gill cover* has a toothed rear edge; *gill cover* has three short spines; *scales* small and continue onto head, lower jaw scaled; *fins* 1 dorsal fin; tail fin rounded. D XI, 13-16; A III, 8-9.

Back green-brown with lighter sides and yellowish underneath. There is a lighter green mottling over the head, back and sides. On the head the spots tend to radiate from the eye and on the body they tend to form vertical bands. The dark fins usually have a lighter edge except the dorsal fin which has an orange edge.

Up to 1.4 m.

These fish are common from 10-400 m. amongst rocks where there are plenty of holes and caves. They are usually solitary and have a well defined home territory which contain a number of refuge holes, often with one or two primary holes which may be occupied by more than one fish. They are not very strong swimmers and spend a considerable amount of time hovering just outside their holes. They are frequently encountered by divers both in holes where they sit slowly moving their fins and in open water where, if not frightened, they frequently turn to face the diver. They spawn during summer possibly at the full moon. They are carnivorous feeding on molluscs, crustacea and, to a lesser extent, fish

Grouper   p. 194 **pl. 25**
Epinephelus alexandrinus Valenciennes

Mediterranean, and neighbouring Atlantic coast as far south as the Cape Verde islands.

Body oval, but more slender than *Epine-*

phelus guaza; jaws, the lower is longer than the upper and very prominent; *front gill cover* has a toothed rear edge; *gill cover* has 3 spines; *fins* 1 dorsal fin. Tail fin either square cut or slightly concave. D XI, 16; A III, 8.

Uniform brown in colour with sides lighter and yellowish underneath. There may be lighter blotches along the sides and back but they are much less distinct than the markings of *Epinephelus guaza*. Young fish have 4-5 distinct dark longitudinal stripes along the body and across the gill covers. These stripes are present in the adult but are much less conspicuous.

Up to 80 cm.

Like *Epinephelus guaza* these fish are common amongst rocks but have a larger and less distinct home territory. They frequently swim in pairs or small groups. Another point of difference with *E. guaza* is that they are stronger swimmers and move greater distances from the rocks into midwater, where they hunt. They are rarely seen hovering around holes. They chiefly feed on fishes. Spawn in summer, possibly at the full moon.

Comber   p. 194
Serranus cabrilla (L.)   **pl. 29, 30**

Mediterranean; East Atlantic north to South-West England; Western Channel; Red Sea.

Body rather long; *teeth* small and pointed and not moveable; *front gill cover* has toothed rear edge; *gill cover* has two spines; *scales* small and continue onto head and lower jaw. Up to 90 scales along the lateral line; *fins* 1 dorsal fin. Tail fin slightly forked. D IX-X, 13-15; A III, 7-8.

Back and sides are reddish brown with 7-9 vertical darker bands. There are 2-3 longitudinal bluish or bluish-green stripes along the head and body. There are also longitudinal yellow stripes. The colour, however, can vary with age, season and depth. Fish that live in the deeper ranges tend to be more red.

Up to 15-20 cm., rarely 40 cm.

Usually found between 20-55 m. but also down to 500 m. over sandy and rocky areas and also over sea grass (*Posidonia*) beds. They feed on fish, crustacea, worms and cephalopods. Spawn from April to July in the Mediterranean and Red Sea and are one of the fishes which are true hermaphrodites, both ovaries and testes mature simultaneously in the same individual.

Painted Comber   p. 194
Serranus scriba (L.)   **pl. 27, 28**

Mediterranean and Black Sea; East Atlantic north to Biscay.

Body oval but less elongate than the Comber (*S. cabrilla*); *front gill cover* toothed on rear edge; *gill cover* with 2 spines; *scales* rather larger than those of *S. cabrilla*; *fins* 1 dorsal fin. D X, 14-16; A III, 7-8.

The back and sides are reddish or yellowish brown with 4-7 dark vertical bands. There is a large pale blue spot on the flanks which prevents confusion with any other species; and a characteristic intricate blue and reddish patterning on the snout and head which resembles arabic characters and gives the fish the name '*scriba*'.

Up to 25 cm.

Found in shallow water usually above 30 m. amongst rocks and areas of sand and sea grass. They are solitary fish and like *Serranus cabrilla* are true hermaphrodites spawning in early summer in the Mediterranean. They are carnivorous and feed upon fish, crustacea and molluscs.

Brown Comber   p. 194
Serranus hepatus (L.)
= *Paracentropristis hepatus* (L.)

Mediterranean; East Atlantic north to Portugal, south to Senegal.

Body oval but rather deeper than either the Comber (*Serranus cabrilla*) or the Painted Comber (*Serranus scriba*); *eye* large nearly equal to the length of the snout; *front gill*

cover toothed along entire edge; *gill cover* with 2 spines; *scales* 50-60 along the lateral line; *fins*, 1 dorsal fin; tail fin rounded. D X, 11-12; A III, 7.

Reddish or yellowish brown with 3-5 vertical dark bands on sides. White on belly and chin. There is a characteristic black spot on the dorsal fin behind the last spine. The pelvic fin is black.

Up to 13 cm.

Found near sand, mud, rocks, and sea grass beds from about 5-100 m. They are carnivorous feeding on small fish, crustacea and molluscs. Like related species this fish is an hermaphrodite spawning in the spring and early summer. Although reputed to be common the authors have never seen this fish and failed to find a photograph of it amongst the hundreds they have examined!

Anthias anthias (L.)   **pl. 31**

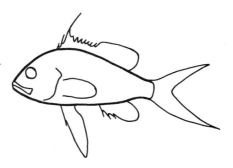

Deeply forked tail fin;
elongate 3rd dorsal spine.
24 cm.

Mediterranean; East Atlantic north to Biscay.

Body oval but very flattened laterally; *profile* steep; *snout* short and rounded; *front gill cover* with a toothed edge; *gill cover* with 2 or 3 spines; *scales* large and continue onto head; *fins* 1 dorsal fin with the 3rd spine greatly elongate. Pelvic fins long; tail fin deeply forked with the lower lobe longer than the upper. D XI, 15; A III, 7.

Back and sides red shading to pink underneath. There are three yellow stripes along the sides of the head which extend onto the front part of the body. Underwater these fish appear grey with long flowing fins. Usually 15-24 cm., rarely longer.

Commonly found below 30 m. amongst rocks where groups, sometimes large, hover just inside the entrances of caves and crannies. They feed on small fish and crustacea and spawn in the spring.

Family:

The **Sea Bream** are mostly found in the warmer seas of the world and only rarely in north European waters. They are exceedingly common in the Mediterranean, especially near the coasts, and any deep-bodied silvery fish seen there is very likely a bream of some kind. Most species form groups or loose shoals but the Saupe swims in such tight and disciplined schools that it is often taken for a sardine—a quite unrelated fish.
They have oval flattened bodies with a moderate or small mouth set low in the head. Both the body and head have distinct, large scales. There is a single dorsal fin, the front part having strong spiny rays, the rear half having soft branched rays. The pelvic fin always has one spiny ray and five soft rays; the anal fin always has three spiny rays but a variable number of soft rays. There are no spines on the gill cover and the pre-operculum is never toothed. The teeth in the jaws are well developed and specialised to serve the particular feeding habits of the fish. Thus there may be chisel-shaped incisors for scraping algae and small animals off rock and weeds, fang-like canines for catching fish or grinding molars for crushing shellfish. There are no teeth on the roof of the mouth.
Their scientific classification relies strongly on their teeth but this is not a helpful character for free-swimming fishes. Most bream can be identified from their markings, especially by their dark spots and longitudinal bands. Some have a characteristic head profile, either very steep, or concave with a protruding mouth. Many appear pink or reddish-brown on the surface, but under water the red tints are not visible and the fish appear a silvery grey.

Sheepshead Bream   pl. 175
Puntazzo puntazzo (Cetti)   p. 200
= *Charax puntazzo* (L.)

Mediterranean and Eastern Atlantic north to Biscay and south to the Congo.

Body oval with an elongated snout; *profile* strongly concave; *mouth* small; *teeth*, there is a single row of forward-sloping chisel-like teeth in each jaw followed by a single row of minute grinding teeth; *fins* D XI, 13-14; A III, 12.
Body silvery-grey with from 7 to 11 dark vertical stripes and a black spot on the tail stalk. There is no yellow on the fins.
Up to 30 cm., rarely 45 cm.

This bream does not form large shoals like most other bream but tends to be more solitary. It lives to a depth of 50 m. amongst algal-covered rocks and reefs or near meadows of sea-grass.
The eggs float and are laid in September and October. By the following spring the young have grown to about 5 cm. and are found in swarms near the coast. It feeds upon algae and also upon the felt of small animals and diatoms that grow on rocks and on algal fronds.
It is chiefly caught in March and October, it is good to eat but not outstandingly so.

Couch's Sea Bream; Pagre   pl. 35
Pagrus pagrus (L.)   p. 200
= *Pagrus vulgaris* Cuvier

Mediterranean, Eastern Atlantic south to Senegal and north to Biscay.

Body oval, *profile* steep with a sharp change in slope above eye; *mouth* terminal and set low in the head; *teeth*, 4-6 strong fang-like teeth in the front of both jaws followed by sharp curved teeth and then by 2 rows of grinding teeth in each jaw; *fins*, D XII, 9-11; A III, 8-9.
Silvery colour with rosy-buff tinted fins, tail fin with a darker rim in the centre often with white tips. In life there is a small redbrown spot at the base of the last ray of the dorsal fin.
Up to 50 cm., rarely 75 cm.

Not very common in our area. In the summer it can be seen at depths greater than 20 m over sand and sea-grass, also near algal-covered rocks. In winter it migrates into the deeper water of the Continental Shelf.
Spawns in the summer, the eggs and larvae are pelagic. At 5-9 mm the larvae have a characteristic bony crest above the eye and a spiny gill cover. At 1.5 cm. the young adopt the adult form and migrate nearer coasts.
It is carnivorous feeding chiefly on crustacea but algal remnants are also found in the stomach.

Two-Banded Bream **pl. 177**
Diplodus vulgaris (Geoffroy) p. 201

Mediterranean and Eastern Atlantic south to Senegal.

Body oval; *profile* slightly concave above eye; *mouth* very small; *teeth*, 8 chisel-shaped forward sloping teeth in the front of each jaw, followed by two rows of grinding teeth; *fins*, D XI-XII, 14-15; A III, 14.
There are two conspicuous black saddle marks, one between the dorsal fin and the eye, the other on the tail stalk. Generally golden brown or grey above and silvery below. There are 15 to 16 longitudinal golden lines on each flank.
Up to 20 cm., rarely 40 cm.

A very common fish especially near algal

covered rocks where there is sand close by. It is found from about 2 to 20 m.

The floating eggs are laid in October and the young live a pelagic life until they are 1 or 2 cm. long when they move close inshore. By late Autumn they have grown to about 3 cm. but do not acquire the black saddles so characteristic of the adult until they are 4 cm. long.
They are chiefly carnivorous feeding on small worms, crustacea etc. They are quite good to eat.

Diplodus cervinus (Lowe) **pl. 178**
= *Diplodus trifasciatus* (Rafinesque) p. 201

Eastern Atlantic north to Spain and south to South Africa. Rarely in Western Mediterranean.

Body oval; *profile* slightly concave; *mouth* small with fleshy lips; *teeth*, in the upper jaw there are 10-12 chisel-like teeth followed by two rows of grinding teeth, in the lower jaw there are 9 chisel-like teeth followed by three rows of grinding teeth on each side; *scales* large, there being only 56-59 along the lateral line; *fins*, D XI-XII, 12-14; A III, 11-12.
There are 4 or 5 very strongly marked dark brown cross-bars extending across the whole flank. The middle cross bars may be divided into two on the lower part of the flank.
Up to 50 cm. perhaps even more.

Very little is known about this fish. It is found near the coast over rocky ground from shallow water to at least 300 m.

Saddled Bream p. 201 **pl. 176**
Oblada melanura (L.)

Mediterranean and Eastern Atlantic from Portugal to Angola.

Body oval; *profile* gently convex; *jaws* equal in length; *teeth*, there are several rows of

Dentex *Dentex dentex* p. 204

Powerful jaw;
large teeth;
blue spots on flanks.

Pandora *Pagellus erythrinus* p. 202

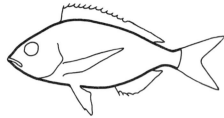

Pointed snout;
pinkish colour, no spots.

Gilthead *Sparus auratus* p. 204

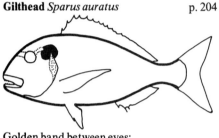

Golden band between eyes;
red spot on gill cover

Spanish Bream *Pagellus acarne* p. 203

Black spot at base of pectoral fin;
depression in dorsal fin.

Couch's Sea Bream *Pagrus pagrus* p. 198

Steep forehead;
brown spot at base of dorsal fin.

Marmora *Lithognathus mormyrus* p. 203

About 12 dark cross bands on body;
long body.

Sheepshead Bream *Puntazzo puntazzo*
p. 198

Elongated snout;
dark cross bands on body;
black spot on tail stalk.

Red Sea Bream *Pagellus bogaraveo* p. 202

Large eye;
often with black spot at beginning of lateral
line.

Annular Bream *Diplodus annularis*   p. 205

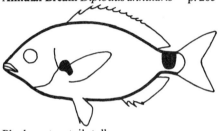

Black spot on tail stalk;
pelvic fin yellow

White Bream *Diplodus sargus*   p. 205

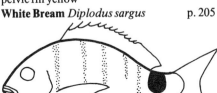

Black spot on tail stalk;
vertical bands on body;
pelvic fin not yellow.

Two-Banded Bream *Diplodus vulgaris*
    p. 199

Two black saddle-like blotches on body.

Diplodus cervinus   p. 199

4-5 conspicuous black cross bands on body;
lips fleshy.

Bogue *Boops boops*   p. 202

Long body;
2 dark parallel stripes along body;
dark lateral line

Saupe *Boops salpa*   p. 203

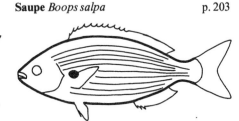

Always in schools;
10-12 golden stripes along body.

Saddled Bream *Oblada melanura*   p. 199

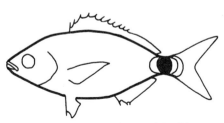

Black spot on tail stalk ringed with white.

Black Bream *Spondyliosoma cantharus*
    p. 204

Blue-black rim on tail;
adult males with humped shoulder.

201

teeth in each jaw. In the outermost row there are incisor-like teeth at the front and sharp conical teeth at the sides. At the front of each jaw are 4 rows of small granular teeth; *fins* D XI, 14; A III, 13-14.

A silvery fish with faint longitudinal dark stippled stripes. On the tail stalk is a characteristic black spot outlined in white along its front and rear margin.
Up to 30 cm.

Although this is a strictly coastal species living near rocks, it is usually seen swimming in small shoals 2-3 m. below the surface in much more open water than most of the other bream. Breeding in late spring the larvae remain in the plankton until late summer. Diet is varied consisting of algae and all manner of small bottom living animals.

Bogue   p. 201 **pl. 37**
Boops boops L.

Mediterranean; East Atlantic south to the Canary Islands and north to Biscay; occasionally in the Irish Sea and North Sea.

Body long; *profile* smoothly convex; *mouth* small, terminal and oblique; *eye* very large; *teeth* small, a single row in each jaw. The upper teeth are notched with 4 sharp points, the lower teeth have 5 points of which the centre one is the largest; *fins* D XIV, 15-16; A III, 13-14.

Blue-green or blue-grey with a yellow tinge. There is a small dark spot at the base of the pectoral fin. The curved lateral line is dark, and when seen in life, there are 2-3 straight dark lines beneath.
Up to 20 cm., sometimes up to 30 cm.

A common species which lives in shoals often in mid-water. They frequently approach rocky coasts or sea grass (*Posidonia*) meadows where they may be as shallow as 2 m. In foul weather, however, they may be found as deep as 150 m.
Spawn in summer, the eggs float and the young feed chiefly on planktonic crustacea.

The young have a characteristic salmon tint. At about 10 cm. they come close inshore. An important food fish in the Mediterranean, the shoals are often attracted to very bright lights carried in small boats by fishermen who then surround them in nets.

Pandora   p. 200
Pagellus erythrinus (L.)

Mediterranean; East Atlantic from northern Biscay to southern Angola.

Body oval; *profile* gently convex, sloping down from the front of the dorsal fin to the snout which is pointed; *nostrils* the two pairs are set close together, the hind ones are enclosed by a flap of skin which is extended forward to protect the front nostrils; *eye* not large; in the adult the eye diameter is less than the length of the space in front of the eye; *teeth*, the teeth in the front of the jaw are small and pointed. Those in the side of the jaw are small and rounded; *fins* D XII, 10; A III, 9-10.

Uniform rosy-red above, paler flanks and silvery below. The gill-flaps have a red rim and the mouth and gill cavities are black.
Up to 25 cm., sometimes up to 60 cm.

Lives near the bottom over sand or mud particularly near rocks at depths ranging from 15-120 m.

Red Sea Beam   p. 200
Pagellus bogaraveo (Brünnich)   **pl. 179**
= *Pagellus centrodontus* (Cuvier)

Eastern Atlantic from Senegal to Ireland, rarely north to Norway and North Sea.

Body oval; *profile* convex with a slight concavity above the eye; *eye* very large, its diameter at least as long as the space in front of the eye; *teeth* in the front of the jaw are short, recurved and sharp, those behind are small and rounded; *fins* D XII-XIII, 11-13; A III, 11-13.

Greyish or rose-tinted. There is a large black spot on the front end of the lateral line beneath the first dorsal ray. In fishes shorter

than 20 cm. this spot may be absent.
Up to 35 cm., sometimes up to 50 cm.

This is a schooling species often found in very large shoals. The younger specimens may enter water as shallow as 40 m. where they favour rocky coasts and wrecks. The larger specimens are more common from about 150 to 500 m.

Although this fish is sufficiently abundant to have some commercial importance it is rarely seen by divers.

Spanish Bream; Axillary Bream   p. 200
Pagellus acarne Risso

Mediterranean and Eastern Atlantic from Senegal to Biscay.

Body long, *profile* convex with a rather blunt snout; *eye* diameter about equal to or slightly less than the length of the space in front of the eye; *teeth*, five comb-like teeth in the front of the jaw, irregular rows of molar-like teeth in the sides of the jaw; *fins* the spiny dorsal rays are of progressively decreasing height after the 2nd spine. D XII, 10-11; A II, 9-10.
Rose coloured above, silvery below with a black spot at the base of the pectoral fin. The mouth cavity is golden or reddish orange.
Up to 25 cm., occasionally up to 35 cm.

Found in small shoals from 20-100 m., although the young fish may be found near the coasts in shallower water. They are mostly caught in springtime over soft bottoms but are rarely seen by divers.

Marmora   pl. 36, 180
Lithognathus mormyrus (L.)   p. 200

Mediterranean and all the warm waters of the Eastern Atlantic.

Body long and laterally flattened; *profile* rather rounded; *jaws* set low in the head and strengthened at the front; *lips* rather thick and wrasse-like; *teeth* there are several rows of fine teeth in the front of the jaw of which

the outer ones are the larger. These are followed by 3-4 rows of molars in the upper jaw, 2-3 rows in the lower; *fins* D XI-XII, 12; A III, 10-11.
Silver with a characteristic pattern of usually 6 strongly marked dark cross bands alternating with an equal number of narrow bands.
Up to 30 cm.

A gregarious species swimming in tightly disciplined schools at depths down to about 20 m. Favours soft bottoms where they feed mostly on small bottom living animals. Particularly common where there are accumulations of decomposing sea grass (*Posidonia*). They fill their mouths with the organic débris, rise half way to the surface, spit out the débris but retain the small crustacea it contains. They are very tolerant of changes in salinity and are able to enter the mouths of rivers as well as to survive in salt lagoons made very saline by evaporation.

Saupe   p. 201 pl. 38
Boops salpa (L.)
= *Sarpa salpa* (L.)

Mediterranean and Eastern Atlantic from Biscay to South Africa.

Body oval and symmetrical; *profile* smoothly convex; *mouth* terminal and small; *teeth*, a single row of cutting teeth in each jaw. The teeth in the upper jaw are notched, those in the lower jaw are triangular and serrated; *fins* D XI-XII, 14-16; A III, 13-15.
Greyish-silver with 10-12 longitudinal golden stripes and a yellow eye. There is a black spot just above the base of the pectoral fin.
Up to 30 cm., sometimes up to 45 cm.
The Saupe is very common in the Mediterranean. Shoals of this species are instantly recognisable by their tight packing and disciplined behaviour. They feed off the algae and associated diatoms that cover rocks and weeds. A shallow-living bream it is never found deeper than 15 m. and the shoals of young may feed in water less than 1 m. deep where they are dragged back and

forth by wave action. Schools of Saupe are often accompanied by other species such as the Amberjack (*Seriola dumerili*) and the Bogue (*Boops boops*).

Spawning is in early spring and late autumn. Fishes of 10-15 cm. are about 1 year old.

Dentex   p. 200
Dentex dentex (L.)   **pl. 34**

Mediterranean and Eastern Atlantic from Senegal to Biscay. Rarely on South coast of Great Britain.

Body oval; *profile* convex and rather steep. Mature males have a fleshy lump over the eyes; *head* massive; *eye* small and set high in the head; *teeth*, there are 4-6 long and well developed canine like teeth in the front of both jaws followed by many smaller teeth of the same shape; *fins* D XI-XIII, 11-12; A III, 7-9.

Colour variable, in specimens up to 1 m. in length the back is bluish-silver above and silvery below with 4-5 indistinct darker cross-bands. The flanks are dotted with small blue spots that loose their colour after death. The pectoral fins have a rosy tint. The very large specimens longer than 1 m. may be a uniform dull red.

Up to at least 1 m.

Lives over rocky ground from about 10 m. to 200 m. In the spring it may come close in to the coasts but in winter retires to deeper water. An active hunter, it feeds on fishes and cephalopods. Dentex is a much prized food fish.

Gilthead   p. 200
Sparus auratus L.

Mediterranean and Eastern Atlantic from north Biscay to Ghana.

Body deep and oval; *profile* rather steep and smoothly convex; *jaws* upper jaw slightly longer than the lower; *lips* are thick; *teeth* 4-6 strong conical teeth in the front of each jaw followed by 4-5 rows of molar-like teeth in the upper jaw and 3-4 rows in the lower jaw; *fins* D XI, 13; A III, 11-12.

Colour variable usually grey or gunmetal above, silvery below. The belly is white. There is a dark blotch at the origin of the lateral line. The edge of the gill cover has a scarlet patch. In life the most characteristic feature is the golden band running between the eyes but the colour fades after death.

Up to 70 cm.

A shallow water species that does not generally go deeper than 30 m. and tolerates, or even prefers, brackish water. Usually found in small groups over mud or sand in the shadow of large rocks, but in the spring they congregate in large numbers where the water is brackish. They remain inshore all summer and in the winter return to deeper water to breed. The Gilthead is very sensitive to cold and may die in cold weather.

Uses its strong jaws to crush crustacea and molluscs. The Romans (who dedicated the fish to Venus) fed it upon oysters which were deemed to improved both its colour and taste.

Black Bream; Old Wife   **pl. 39, 40, 41**
Spondyliosoma cantharus (L.)   p. 201
= *Cantharus lineatus* Günther

Mediterranean; Eastern Atlantic north to Scotland and south to Senegal.

Body elliptical and rather deep. The mature males have a humped shoulder and concave forehead. The young, 10-15 cm. in length have a convex forehead and a sharp snout; *mouth* terminal and small set rather low, not reaching back to the level of the eye; *teeth* there is a single row of slightly curved small pointed teeth in each jaw. The teeth in the front of the jaws are larger than the rest; *fins* D XI, 12-13; A III, 9-11.

The colouration of this fish is very variable, the young up to about 20 cm. have a grey or yellowish back with silvery flanks bearing numerous longitudinal light stippled stripes. The fins are sometimes spotted with white and there is a broad dark rim on the tail fin. The adults are dark grey or blue-grey sometimes very dark. There may be 6-9 lighter vertical cross bars on the sides. The spawning males have an irredescent blue-grey band between the eyes. On the nest the male is almost black.

Up to 40 cm., sometimes up to 50 cm.

In the Mediterranean it is generally found over sand and sea grass (*Posidonia*) meadows at depths greater than 15 m. In northern Europe it is commonly fished from around wrecks and rock areas and is frequently seen by divers. This is one of the few bream that lays its eggs on the bottom, generally in early summer. The male follows the female and digs an oval pit in the sand with his tail and the female deposits her slightly sticky eggs into it in a single layer. The young keep to the vicinity of the nest until they are about 7-8 cm. long when they leave the coast for deeper water.

Annular Bream   p. 201
Diplodus annularis (L.)

Mediterranean and Eastern Atlantic south to Senegal.

Body oval and flattened; *profile* slightly convex; *jaws* equal in length; *teeth* prominent and chisel-shaped in the front of each jaw with robust rounded grinding teeth behind; *fins*, D XI, 12-13; A III, 10-11.

Greyish or brownish silver with a dark spot on the tail stalk and another at the base of the pectoral fin. There may be 4 or 5 darker crossbands on the flanks. The pelvic fins are yellow.

Up to 12 cm, rarely 20 cm.

One of the commonest of the Mediterranean bream it normally shoals around rocky coasts especially near deep water.

Family:

CENTRACANTHIDAE

Closely resemble the Sea Bream (Sparidae) but the teeth are minute and the mouth is protrusible. There is usually a rectangular black spot on the centre of the flank.

It is also seen searching for food amongst sea grass. It is found throughout the year and in spring may enter brackish water.

The eggs float and are laid in April and June. The young, from 9-15 cm in length, may be found some kilometers from the shore amongst floating seaweed etc. but when they reach a length of 3 cm they come close inshore.

Feeds on small worms, crustacea etc.

White Bream   p. 201 **pl. 174**
Diplodus sargus (L.)

Mediterranean and Eastern Atlantic south to Angola and north to Biscay.

Body oval; *profile* gently convex; *mouth* rather small; *teeth*, 8 chisel-like teeth in the front of each jaw with rounded grinding teeth behind; *fins*, D XI-XII, 12-15; A III, 13-14.

Silver-grey with 7-8 dark vertical bands and a dark rim on the tail. The pelvic fin is grey. There is a dark spot on the tail stalk. Up to 45 cm.

This bream is common near rocky coasts especially where fallen rocks form a slope containing many holes suitable for refuge. Found all the year round but is especially common in spring and summer. It is usually found between about 2 and 20 m. The White bream often forms shoals although the larger individuals are usually solitary. In early summer it may enter brackish water, but returns to the sea in the autumn.

The eggs which float are laid in April or June and the young fish live a pelagic life for the remainder of the summer.

It feeds on crustacea, molluscs and echinoderms which it crunches up with its strong grinding teeth.

Picarel *Maena smaris*

Rectangular spot on body;
long body;
sharp snout;
dorsal fin of uniform height.
20 cm.

Picarel
Maena chryselis

Rectangular spot on body;
dorsal fin in 2 lobes.
21 cm.

Blotched Picarel *Maena maena*   p. 205

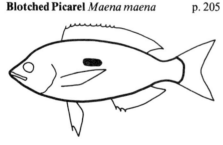

Rectangular spot on body;
distinct shoulder;
deep body;
dorsal fin of uniform height.
24 cm.

Blotched Picarel   pl. 43
Maena maena L.   p. 206

Mediterranean except Black Sea; Eastern Atlantic south to Canary Isles and north to Portugal.

Body rather deep, the length from snout to the rear edge of the gill cover less than greatest height of body. The proportions of the body varies with age and sex; *profile* concave above the head; *scales* 70-75 rows of scales along the lateral line; *fins* dorsal fin of uniform height throughout its length. D XI, 12; A III, 10.

Colour varies with age, sex and season. Always with a rectangular black patch on the flanks, 2 scales high and 7-8 scales long in the young, 3 scales high and 6-7 scales long in the adults. Often blue-grey above, silvery below. The dorsal, anal and tail fins are brownish blue with irregular blue spots. The pectoral fin is brownish-yellow. Adult males may have brilliant blue bands on the head. Female up to 21 cm., male up to 24 cm.

A non-migratory fish which tends to swim in mid-water. Spawns in late summer at about 10-20 m. depth. The male excavates a circular nest in the sand where the female deposits her sticky eggs. These are immediately fertilized by the male.

Picarel   pl. 42
Maena chryselis Valenciennes

Mediterranean.

Body oval, length between the snout and the rear of the gill cover equals the maximum height of the body; *profile* almost straight above the eyes; *scales*, there are 70-80 scales along the lateral line; *fins* maximum height of spiny dorsal rays greater than maximum height of the soft dorsal rays. There is a distinct dip in the middle of the dorsal fin. D XI, 11-12; A III, 9-10.

Colour varies with age, sex and season. Generally bluish-grey above and silver below with a pronounced rectangular spot in the centre of the flank. There is a brownish spot at the base of the pectoral fin. The fins are generally greyish or orange-yellow. Females up to 18 cm., males up to 21 cm.

A non-migratory gregarious fish that is usually found from 70-130 m. They appear to prefer muddy bottoms. Spawn in spring.

Picarel
Maena smaris (L.)

Mediterranean and Eastern Atlantic from the Canary Islands to Biscay.

Body rather long; the length from the snout to the rear edge of the gill cover more or less equals the maximum height of the body; *profile* straight or slightly concave above eye, the snout is pointed; *scales* 80-94 along the lateral line; *jaws*, the upper jaw slightly longer than the lower; *fins* the dorsal fin is of equal height throughout its length but varies in proportion with sex and age. D XI, 11-12; A III, 8-10.

Colour varies with age and season. Generally bluish or brownish-grey above, silvery below. Longitudinal lines of blue spots run along the flanks. There is a rectangular spot in the middle of the flanks. The fins are yellowish or grey. In the male there is a single row of blue spots along the centre of the anal fin and in a band across the tail.

Females up to 15 cm., males up to 20 cm.

A gregarious fish living in 50-100 m. but not close to the bottom. In late winter they gather in huge shoals to breed.

Family:

MULLIDAE

Mullus is the only genus of this numerous tropical family which is native to our area, but other members of the family have penetrated into the Mediterranean through the Suez Canal.

There are two sensory barbels on the chin; the forehead is steep; the body is covered with large scales and there are two well-separated dorsal fins. Members of this family always live near the bottom.

Red Mullet *Mullus surmuletus*

Sloping profile;
2 scales on cheek.
40 cm.

Red Mullet *Mullus barbatus*   p. 208

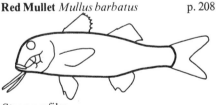

Steep profile;
3 scales on cheeks.
40 cm.

Goat Fish *Pseudupeneus barberinus* p. 208

Two barbels on chin;
long black stripe;

Red Mullet   pl. 44, 45
Mullus surmuletus L.

Mediterranean and Eastern Atlantic from Scotland south to the Canaries.

Body rather long and flattened; *profile* steep; *barbel*, there is a pair of sensory barbels beneath the jaw; *scales* large, there are two large scales on the cheek beneath the eye; *fins* 1D VII-VIII; 2D 8-9; A II, 6-7.

Colour varies with depth, emotion and time of day. In the Mediterranean fishes living shallower than about 15 m. are basically yellow-brown, those living deeper are red.

In both cases the scales have dark edges. During the daytime, especially when the fishes are shoaling they have a pronounced dark red or brown longitudinal stripe from the eye to the tail and 4 or 5 longitudinal yellow stripes. At night the pattern breaks up into an indistinct marbled pattern.
Up to 20 cm., rarely 40 cm.

Swim either singly or in groups of up to about 50. The young browse on algal-covered rocks as shallow as 1 m. but the adults live over sand and mud from about 3 m. down to about 90 m. They feed on small animals buried in the sand which they locate with their sensory barbels and may excavate holes as deep as themselves in their search for food.

Small wrasse and other species often follow behind them when they feed in this way for any morsels of food that are thrown up in the process. When resting on the bottom they can be approached quite closely but when the diver comes too near the 1st dorsal fin is erected which appears to be a signal to the other mullets that it is time to swim off.
Spawns in summer, the female deposits her eggs on a mud or sand bottom at depths from 10-55 m. The young live in open water and are blue in colour but when they reach about 5 cm. they take up a bottom-living existence.
The Red Mullet has been a highly prized food fish since ancient times. The Romans indeed would pay more for a good specimen than they would for the fisherman who caught it. A mullet in a bowl was also used as a pre-banquet entertainment so that the guests might marvel at the rapid changes in colour of the dying fish.

Family:

SCIAENIDAE

Red Mullet
Mullus barbatus L.

There is still argument amongst specialists as to whether *Mullus barbatus* and *Mullus surmuletus* are the same species or not.

'*Mullus barbatus*' has a steeper profile, 3 scales beneath the eye and the longitudinal lines are not present.
Generally lives deeper than *Mullus surmuletus*, and may be caught from 300 m. It is an important bottom fish of the Mediterranean coast of Israel but in the Eastern Atlantic it probably does not occur further north than Biscay.

Goat Fish   p. 207 **pl. 46**
Pseudupeneus barberinus (Lacépède)

Entire tropical Indo-Pacific, Red Sea and eastern Mediterranean.

Resembles the Red Mullet *Mullus surmuletus* in general form except that the profile slopes more gently. *Scales* there are 28-31 along the lateral line. 1D VII; 2D I, 8; A I, 6.
Red-brown or rose colour with a dark longitudinal stripe running from the snout through the eye to the upper part of the tail stalk: however this stripe may be shorter. There is a dark spot at the base of the tail stalk.
Up to 60 cm.
This is one of those fishes that has migrated through the Suez Canal into the Eastern Mediterranean where it is now established.
Generally in small groups on sandy bottoms where they forage for small bottom-living animals. Probably not deeper than 30 m.

Perciform fishes with a wide tail stalk, two dorsal fins and a lateral line that extends onto the tail fin. The otoliths are very large. The swim bladder is well-developed and can be caused to oscillate by the vibration of special muscles. This makes the loud grunting noise for which many of these fishes are famous.

Meagre *Argyrosomus regius*

Long slender body;
silvery in colour.
2 m.

Brown Meagre *Sciaena umbra*

Heavy body;
dark bronze colour;
conspicuous white spines on anal and pelvic fins.
40 cm.

Corb *Umbrina cirrosa*   p. 210

Diagonal golden striping edged with brown-violet;
1 m.

Meagre

Argyrosomus regius (Asso)
= *Sciaena aquila* Lacépède
= *Johnius hololepidotus* Lacépède

Mediterranean; East Atlantic north to Britain, south to Guinea; English Channel; occasionally North Sea.

Body long and slender; *snout* rounded; *mouth* long; *jaws* of equal length or lower jaw slightly longer than the upper; *scales* large on body, smaller on head, between 50-55 along the lateral line. The scales appear to run in oblique rows; *lateral line* continues onto tail; *barbels* none; *fins*, 2 dorsal fins distinct from each other but not separated by a space. Anal fin short. 1D IX-X; 2 D I, 27-29; A II, 7-8. *Swim bladder* very large, it occupies nearly the whole of the abdominal cavity.

Back silvery grey or brown, sides lighter with golden and silver reflections. Under surface silver. The fins are darker, grey or brown. The inside of the mouth is golden. Sometimes there is a darker spot on the gill cover. Up to 200 cm.

Found in shallow water and is able to withstand brackish water. Young fish particularly are found in estuaries. Usually seen amongst rocks. They are carnivorous and hunt shoals of small fish. The large swim bladder is used to produce deep sounds which may be heard over several metres. Day (1880) reports that the Dutch were said to perceive an image or representation of the Virgin on each scale.

Brown Meagre; Corb   **pl. 47**

Sciaena umbra (L.)
= *Johnius umbra* (L.)
= *Corvina nigra* Valenciennes

Mediterranean and Black Sea; East Atlantic from Gulf of Gascony to Senegal.

Body deep and large; *profile* curved; *snout* rounded; *jaws* large extending to the rear edge of the eye; *scales* fairly large; *fins*, 2 dorsal fins distinct but joined by a fine membrane. Anal fin short the 2nd spine large. Tail fin square cut in the adult, slightly indented in the young. 1D X; 2D I, 23-25; A II. 7-8

Back and sides brown-bronze with golden reflections. Under surface lighter and silvery. The fins are dark but the spines of the

pelvic and anal fins are white and appear very conspicuous underwater.
Up to 40 cm., sometimes up to 70 cm.

These fish are found in small shoals from 5-20 m. They are seen amongst rocks often in caves and cracks and amongst sea grass (*Posidonia*) where they sit quietly with only their tails slowly moving. Spawning in the Mediterranean occurs in late spring and summer. Carnivorous, they feed on fish, crustacea and molluscs.

Corb   p. 209
Umbrina cirrosa (L.)

Mediterranean and Black Sea; East Atlantic north to Biscay, south to Senegal; Red Sea.

Body elongate and laterally flattened; *pro-file* gently curved; *snout* rounded; *jaws*, upper jaw longer than lower; *barbels*, there is a small fleshy barbel at the extreme tip of the lower jaw; *scales* 50-60 along the lateral line; *lateral line* follows the curve of the back and extends onto the tail; *fins* 2 dorsal fins distinct but not separated from each other. Tail fin slightly concave in upper half. 1D X; 2D I, 21-25; A II, 7.
Back and sides silvery with diagonal golden stripes edged with fine brown-violet stripes. Under surface silvery. The rear edge of the gill cover is black.
Up to 1 m.

These fish are solitary and found over mud and sand and in rocky areas. Young fish may be seen in estuaries. They are carnivorous and feed on molluscs, worms and crustacea. Spawning occurs in June in the Mediterranean.

Family:

POMATOMIDAE

Ferocious fishes hunting in schools in all the warm waters of the world. There are two dorsal fins of which the first is lower than the second.

Blue Fish   **pl. 181**
Pomatomus saltator (L.)

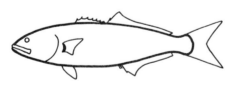

2nd dorsal fin taller than 1st dorsal;
2 spines before anal fin.
60 cm.

Mediterranean and in all warm seas.

Body rather long and laterally flattened; *mouth* large extending back behind the eye; *jaws*, lower jaw longer than upper; *scales* very small, 95-106 along the lateral line; *fins* 2 dorsal fins, the 2nd being much taller than the first. The anal fin is preceded by two short spines. 1D VII-VIII; 2D I, 25-28; A II + I, 25-27.
Blue-green above with silver flanks and a whitish belly. There is a black spot at the base of the pectoral fin.
Up to 60 cm., occasionally up to 100 cm.

A ferocious and swift predator living in large shoals in open water. They come into coastal water and will attack fishes almost as long as themselves. A hunting school will leave a trail of blood and maimed fishes behind it.

210

Family:

CARANGIDAE

The Horse and Jack Mackerels mostly live near the surface in the open sea. Most are swift swimmers and voracious hunters. Superficially they resemble the Mackerels and Tunas but have no finlets behind the dorsal and anal fins. There are usually two short spines in front of the anal fin and in some species the lateral line is armoured with keeled scales. The tail stalk is narrow.

Horse Mackerel; Scad   **pl. 48, 49, 50**
Trachurus trachurus (L.)   p. 212

Mediterranean and Black Sea; Atlantic; English Channel; North Sea; West Baltic.

Body long and slender; *snout* pointed; *mouth* large and slanted downwards; *jaws* prominant extending to beneath the eye; *eyes* large with transparent eyelids at each side; *scales* small and easily rubbed off; *lateral line* has a row of 69-79 wide bony scales each with a point on their rear edge. The lateral line follows the back until the 2nd dorsal fin where it dips downwards. There is a secondary sensory canal which runs from in front of the 1st dorsal fin to near the rear end of the 2nd dorsal fin; *fins* 2 dorsal fins, 1st consists of rays and is higher than the 2nd. 2nd dorsal fin long. Anal fin long and has 2 isolated spines situated in front of it. Pectoral fin long reaching back to the anal fin. Tail fin forked. 1D I-VIII; 2D I, 28-33; A II + I, 25-33.
The back is grey or bluish-green. The flanks are silvery with metallic reflections which are lost after death. Under surface silver. The gill cover has a small black spot. Young fish are completely silver.
Up to 40 cm.

Horse Mackerel are common fish found swimming in shoals, sometimes very large, from 10-100 m. usually in open water. During the summer months they are common near the coasts but during the winter they migrate offshore to deep water sometimes below 500 m. They are particularly common in shallow sandy areas particularly off-shore sand banks like those in the North Sea and off the Dutch coast. They are carnivorous. Adult fish feed upon small shoals of fish (herring, sprats etc.), crustacea and cephalopods. Young fish feed on minute larvae and crustacea present in the plankton. Spawning occurs during the summer. Young fish are frequently seen in small groups swimming around the tentacles of jellyfish or around floating debris. They are not considered an important food fish but they are processed for fish meal and are the subject of fairly extensive fisheries off Spain and Portugal.

Horse Mackerel   p. 212
Trachurus mediterraneus (Steindachner)
= *Caranx trachurus* Steindachner

Very similar to *Trachurus trachurus* except that the body is more flattened laterally, the scales along the lateral line are smaller (between 78-92) and the secondary sensory canal ends under the 3rd or 4th ray of the 2nd dorsal fin. The last rays of both the 2nd dorsal and anal fins end in small distinct membranes. 1D I-VIII; 2D I, 32-33 + 1; A II + 1, 25 + 1.

Amberjack; Yellow Tail   p. 212 **pl. 51**
Seriola dumerilii (Risso)

Mediterranean; southern Atlantic.

Body long and laterally flattened; *profile* gently curved; *snout* rounded; *jaws* extend to middle of eye; *scales* very small 150-180 along lateral line; *fins*, 2 dorsal fins, 1st much smaller than the 2nd. Anal fin is

Horse Mackerel *Trachurus trachurus*

p. 212

69-79 large scales along lateral line;
2nd sensory canal along back ends near rear
of 2nd dorsal fin.
40 cm.

Horse Mackerel *Trachurus mediterraneus*

p. 211

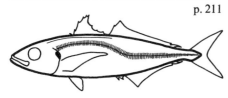

78-92 scales along the lateral line;
2nd sensory canal along back ends under
the 3rd or 4th ray of 2nd dorsal fin.
40 cm.

Amberjack *Seriola dumerilii*   p. 211

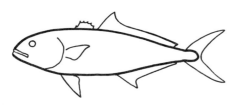

No pre-anal fins or spines;
yellowish sides.
2 m.

Pilot Fish *Naucrates ductor*   p. 213

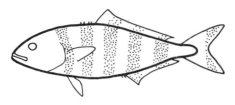

5-7 dark vertical stripes.
35 cm.

Leer Fish *Lichia amia*   p. 213

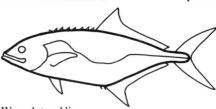

Wavy lateral line.
1 m.

Campogramma vadigo   p. 213

Straight lateral line;
the greyish-green back ends in a deeply indented line along the sides.
65 cm.

Pompano *Trachinotus glaucus*   p. 213

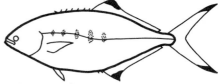

4-6 spots on sides.
30 cm.

shorter but similar to the 2nd dorsal. Tail
fin deeply forked. 1D VI-VIII; 2D 34-39;
A II, 18-20.
Back silvery blue or grey; sides lighter and
under surface silver. The flanks have a gold-
en iridescence and young fish have a distinct yellow eye. There is a diffuse dark
stripe running from the shoulder to the eye.
Young fish are yellow with dark vertical
stripes.
Up to 2 m.

Found in small fast-swimming groups usually around rocks from moderate to deep
water. Young fish are sometimes found

underneath the bells of jellyfish and occasionally amongst shoals of Saupe (*Boops salpa*). They are carnivorous and spawn in spring and summer.

Pilot Fish   p. 212
Naucrates ductor (L.)   **pl. 182**

Mediterranean; English Channel; East Atlantic and all warm and temperate oceans.

Body elongate; *profile* curved; *snout* rounded; *scales* very small and extend onto cheeks; *lateral line* has a keel at each side of the tail stalk; *fins*, 2 dorsal fins the 1st present as 3-5 short separate spines situated in front of the 2nd dorsal which is long. Anal fin has 2 separate spines situated in front of it. Tail fin forked. 1D IV-V; 2D I, 26-28; A II + I, 16-17.

Bluish with 5-7 blue-black, grey or brown-black dark vertical stripes which extend onto the dorsal and anal fins. There is also a dark stripe near the end of the tail fin which also has a white edge.
Up to 35 cm., sometimes up to 70 cm.

An open water fish which is frequently seen in association with sharks, mantas, turtles, boats and driftwood. The young in particular congregate in small groups under and in the neighbourhood of large jellyfish, pieces of wreckage and floating weed. The reasons for this association are not known. Often found in small groups. They are carnivorous and feed mainly on planktonic crustacea and occasionally fish and molluscs.

Leer Fish   p. 212
Lichia amia (L.)

Mediterranean; East Atlantic from Portugal to South Africa.

Body deep and laterally flattened; *snout* pointed; *jaws* long extending to the rear edge of the eye; *scales* small, do not extend onto head; *lateral line* wavy; *fins* 2 dorsal fins, 1st consists of 7 small separate spines each with a membrane on their rear edge.

The 1st spine is directed forward. 2nd dorsal and anal fins are similar with front part elongate. Tail fin deeply forked. 1D VII; 2D I, 20-21; A II + I, 20-21.

Back bluish or greenish white or greyish. Sides and belly silvery. The fins are dark and the lateral line is black. Young fish have several more or less distinct cross bands, sometimes replaced by dark spots.
Up to 1 m.

A mid-water fish which occasionally approaches shore, particularly in the spring to breed although they also pursue shoals of fish into shallow water. Little else is known except that they feed mainly on fish.

Campogramma vadigo (Risso)   p. 212
= *Lichia vadigo* (Risso)

Mediterranean and adjacent East Atlantic.
Body oval, less deep than the Leer Fish (*Lichia amia*) and the Pompano (*Trachinotus glauca*); *snout* rounded; *jaws* long extending behind the rear edge of the eye; *eye* small; *lateral line* slightly wavy in front half and straight in second half; *fins* 2 dorsal fins, 1st consists of 7 separate spines; 2nd dorsal long with elongate rays. Anal fin similar to 2nd dorsal but shorter. Tail fin forked. 1D VII; 2D I, 19-31; A II + I, 23-24

Back a greyish green which continues onto the sides where it changes into silver with a more or less distinct wavy line with 15-20 indentations along it. Sides and belly silvery white. Fins greyish.
Up to 65 cm.

Little is known about the habits of these fish except that they live in mid-water are carnivorous and feed on fish, molluscs and cephalopods.

Pompano; Glaucus; Blane   **pl. 52**
Trachinotus glaucus (L.)   p. 212

Mediterranean; southern Atlantic north to the English Channel, occasionally north to Scandanavia.

Body oval, laterally flattened; *snout* short and rounded; *mouth* small reaching just beyond the front edge of the eye; *scales* small about 127 along the lateral line; *fins* 2 dorsal fins, 1st fin consists of 5-7 separate spines the 1st being directed forward. 2nd fin long and similar to the anal fin. Anal fin has 2 separate spines in front of it. Tail fin forked. 1D I + V-VI; 2D I, 24-25; A II + I, 23-25.
Back bluish or greyish silver, sides and belly pinkish silver. The sides may have a yellowish tinge. There are 4-6 dark oval spots along the sides. The tips of the 2nd dorsal, anal and tail fins are black. The 2nd dorsal and anal fins are yellowish.
Up to 30 cm., occasionally 50 cm.

Mid-water migratory fish which occasionally approaches shore in small groups. They breed during summer and feed on small fish.

Family:

CORYPHAENIDAE

Only one species in this family. The body is long and compressed. The dorsal fin begins on the head and extends to the tail. Teeth very small.

Dolphin Fish   pl. 191
Coryphaena hippurus L.

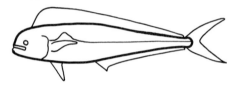

Long flattened body;
long dorsal fin;
steep forehead.
190 cm.

Mediterranean and all warm seas.
Body elongate and very compressed laterally; *profile*, in the young fish rounded but with increasing age it becomes progressively steeper and the adult males have a forehead that is almost vertical; *scales* minute, embedded in the skin and not visible to the naked eye; *lateral line* is straight except over the pectoral fin where it is humped; *jaws*, lower jaw protrudes beyond the upper; *fins*, there is a single dorsal fin running the length of the body from the eye to the tail fin which is deeply forked. D 55-65; A I, 26-30.
Blue-green above, silvery below with brilliant silver and gold iridescence. Stippled with dark and gold spots.
Up to 190 cm.

A free swimming species of open water which undertakes long and regular breeding migrations. They are often associated with other species such as Pilot Fish and seem to like the shade cast by flotsam and small boats. Feed on fishes and may be caught by trolling. At night they are one of those fishes attracted to a bright light. Very good to eat.

Family:

BRAMIDAE

Deep bodied oval fishes with strong persistant scales which have a longitudinal ridge.

Ray's Bream
Brama brama (Bonaterre)
= *Brama raii* (Bloch)

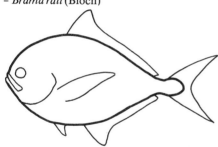

Very deep nearly round body;
fins not elongate.
70 cm.

Mediterranean; East Atlantic from Scandanavia to Madeira; also Pacific from Alaska to southern California; English Channel; North Sea.

Body very deep and very flattened laterally; *profile* very steep and curved; *snout* rounded; *mouth* short and directed steeply downwards; *eye* situated along the central axis of the body; *scales* cover whole surface except the snout and they also continue onto the fin rays. 70-90 scales along the lateral line; *fins* 1 dorsal fin, the front rays longer than the others. The anal fin is similar but shorter than the dorsal. Pectoral fin very long. Tail fin deeply forked. D III-V, 30-33; A II-III, 27-28.

Back greyish, greenish-brown or even bluish. The sides and under surface are silver and the whole fish gives a silver appearance. The pectoral fins and the area around the eyes are golden. The edges of the dorsal and anal fins are dark. The silvery colour fades rapidly after death when the fish appears a uniform dark grey.
Up to 70 cm.

Found in open water from depths ranging from 100-400 m. and may sometimes be found washed up on shore. In the nothern part of their distribution (Biscay northwards) they are not common and solitary specimens are usually encountered in winter; in some years, however, they become quite numerous.

They feed mainly on small shoaling fishes and cephalopods. Spawning occurs in deep water at a water temperature greater than 20°C and young fish have been found in the Mediterranean at depths below 2700 m. and also in mid Atlantic waters.

Family:

CEPOLIDAE

The body is very long and ribbon shaped. The dorsal and anal fins are very long (the dorsal extends the entire length of the body) and both are joined to the tail fin. No spiny fin rays.

Red Band Fish; Red Snake Fish   pl. 53
Cepola rubescens L.

Long, slender, flattened body;
colour reddish.
50 cm.

Mediterranean; East Atlantic from Britain to Senegal; Western Channel.

Body very long and slender gradually tapers to the tail; laterally flattened; *snout* blunt; *mouth* large and directed downwards; *teeth* similar in each jaw, they are pointed and curved inwards; *eye* large diameter equal 1/3 head length; *scales* very small; *fins* one dorsal fin which extends the whole length of the body and one anal fin which is shorter. The tail fin is slender and the rays are extended into short filaments. D 67-69; A 56-62.

The back is reddish and the sides are lighter

215

with a silver sheen and the under surface is orange or yellow. The dorsal fin is yellow with a large red area near the front of the fin. Anal fin yellow, pectoral fin pink, tail fin reddish.

Up to 70 cm., commonly between 30-50 cm.

Found from 70-200 m. usually in mud into

which they are thought to burrow. They have, however, been caught from amongst sea grass and from rocky areas. Sometimes they may be caught from shallow water particularly after storms. Although much of their time is spent on the sea bed they do leave it to swim in mid-water where they feed on the small planktonic crustacea.

Family:

POMACENTRIDAE

There are many species of Pomacentrid fishes in tropical waters but there is only one representative in our area.

The body is small, rather deep and laterally flattened. The mouth is small. There is a single dorsal fin having a long section of spiny rays and a short section of soft rays. The soft rays, however, are taller than the spiny ones.

Damsel Fish pl. 55, 56, 57
Chromis chromis (L.)

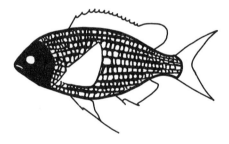

Adults very dark colour;
forked tail;
young brilliant iridescent blue.
15 cm.

Mediterranean and Eastern Atlantic from Portugal to Guinea.

Body oval and laterally flattened; *mouth* small, terminal and oblique; *scales* very large, 24-30 along the lateral line; *fins*, a single dorsal fin with two distinct lobes, the soft rays being the longer. D XIV, 10-11; A II, 10-12.

The adults are dark brown, the scales being

outlined in darker brown. The young up to about 10 mm. have brilliant iridescent blue stripes running along the head and flanks. The blue colour gradually disappears as the fish gets bigger and the adults have no blue colour.

Up to 15 cm.

One of the most common fishes in the Mediterranean. They are generally seen in huge almost stationary shoals near rocky coasts in mid-water. Do not usually go deeper than 25 m.

Spawning is in summer. Mixed shoals choose a spawning site on a rocky slope where there are flat depressions or large rocks; preferably in depths from 2-15 m. Occasionally they will choose a sandy area in deeper water. The male appears to attract the female by fanning his tail and making jumping movements. The females hover in a swarm nearby occasionally descending to the spawning places. Fertilization takes place after the eggs are laid. After three days the females leave the area but the males continue to guard the spawn.

The brilliant blue young hover in depressions and sheltered crannies in the rocks and are a common sight in late summer.

Family:

The **Wrasse** are extremely common in shallow water both in the Mediterranean and in northern European waters. Some possess the most brilliant colours of any fish in our seas. Some can easily be identified under water, but others are very difficult. This is partly because the sexes can be strikingly different in colour and in the breeding season the males especially may become much more vividly patterned. In addition there are colour changes associated, amongst other things, with their emotional state, the season and the type of bottom. Many Wrasse are known to change their sex but it is more doubtful if any can function as male and female at the same time.

Wrasse often show elaborate courtship and nesting behaviour. Some wrasse notably *Crenilabrus melanocercus* in the Mediterranean act as 'cleaners'. That is they feed from the fungus and animal parasies that infest the skin of the host fishes. Definite cleaning stations exist—generally near some bold feature of the bottom like a large rock. The fish to be cleaned invite attention by hanging stationary in the water with fins spread and gill covers gaped. The young of some Wrasse e.g. *Coris julis* will act as cleaners although the habit is lost in adults.

The Wrasse are mostly carnivorous feeding on small bottom living animals or tiny morsels in the water. Most, perhaps all, are only active during the day and take refuge in rock crevices, or perhaps bury themselves in the sand at night.

The Labridae have a terminal mouth with fleshy lips. The teeth are well developed. The main gill cover is not spined although the front gill cover may be finely toothed. The scales are large and easily visible. The lateral line is complete. The single dorsal fin is distinctly divided into a hard and soft-rayed portion.

Cuckoo Wrasse pl. 58, 59
Labrus mixtus L. p. 218
= *Labrus bimaculatus* (L.)

East Atlantic from Senegal to Northern Scotland. Mediterranean.

Body rather elongate; *mouth* reaches back almost to the eye and the lips are thick; *teeth* large and conical arranged in a single row the largest being in front; *front gill cover* smooth; *lateral line* parallel to the back; *fins*, dorsal of almost equal height throughout its length. D XVI-XIX, 11-14; A III, 9-12.

In the mature fish the sexes differ greatly in colour and pattern. The females and immature males are yellow or reddish-orange with 3 dark blotches on the back, 2 below the soft dorsal rays, the 3rd on the tail stalk. The dark markings are often spaced with lighter markings making the pattern very conspicuous. The males have brilliant blue heads with darker horizontal bands extending onto the flanks which are yellow or orange. There are no blotches on the back of the male. During courtship the males have a white patch on the forehead.
Up to 35 cm.

Common during the summer around rocks at depths greater than about 10 m. The 'nest' is a depression in the gravel bottom and the male drives off other males that come too close. The eggs are laid during the summer. The younger fish tending to live in the shallower water. Their food appears to be chiefly bottom living animals such as crabs and whelks. A long-lived fish; a specimen of 28 cm. may be 17 years old. The older females may take on the patterns and colour of the adult male and it is possible that they achieve the full sexual function of the male as well.

Cuckoo Wrasse *Labrus mixtus*   p. 217

Adult female and young: 3 dark blotches
on rear back.
35 cm.

Cuckoo Wrasse *Labrus mixtus*   p. 217

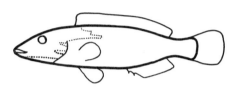

Adult male: blue head.
35 cm.

Green Wrasse *Labrus turdus*   p. 220

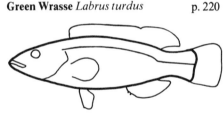

Long body;
step in lateral line.
45 cm.

Brown Wrasse *Labrus merula*   p. 221

Mouth very small;
dorsal, anal and tail fins with blue rims.
45 cm.

Ballan Wrasse *Labrus bergylta*   p. 221

Scales usually with a light centre and
darker rim;
young fish usually green;
front gill cover not toothed.
40 cm.

Axillary Wrasse *Crenilabrus mediterraneus*
      p. 221

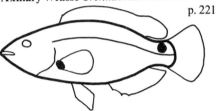

Black spot at base of tail stalk above lateral
line;
dark spot at base of pectorals.
15 cm.

Black-tailed Wrasse   p. 222
Crenilabrus melanocercus

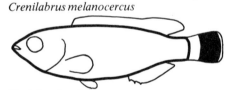

Dark band on tail;
Mediterranean;
3 anal fin spines.
14 cm.

Corkwing Wrasse *Crenilabrus melops*
      p. 222

Black spot on tail stalk on or below lateral
line sometimes obscured in adult male.
6 m.

Ocellated Wrasse *Crenilabrus ocellatus*

p. 222

Spot outlined in red on gill cover;
black spot on base of tail stalk.
13 cm.

Grey Wrasse *Crenilabrus cinereus*   p. 224

Lateral line higher at rear than at front;
dark spot set low on tail stalk.
8 cm.

Painted Wrasse *Crenilabrus tinca*   p. 223

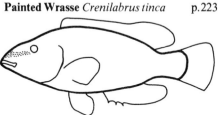

Slight hump on snout;
dark band between eyes.
30 cm.

Long-snouted Wrasse *Crenilabrus rostratus*

p. 224

Very long snout;
concave profile.
12 cm.

Five-spotted Wrasse   p. 223
Crenilabrus quinquemaculatus

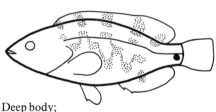

Deep body;
large scales;
often a black spot on tail stalk below lateral
line.
15 cm.

Rock Cock *Centrolabrus exoletus*   p. 224

Black bar on tail;
North East Atlantic;
IV-VI anal fin spines.
15 cm.

Crenilabrus doderleini   p. 224

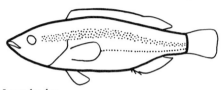

Long body;
dark and light longitudinal band along
body;
10 cm.

Goldsinny *Ctenolabrus rupestris*   p. 225

Red-brown colour;
black spot high on tail stalk.
18 cm.

Lappanella fasciata   p. 225

Black spot on soft rays of dorsal fin;
black spot on tail fin.
Up to 14 cm.

Rainbow Wrasse *Coris julis*   p. 225

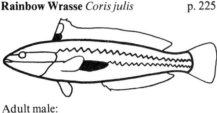

Adult male:
long body;
black and orange spot on elongate dorsal
rays;
zig-zag orange stripe on flanks.
25 cm.

Rainbow Wrasse *Coris julis*   p. 225

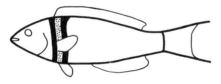

Female and young:
long body;
blue spot on gill cover;
light stripe along flanks.
18 cm.

Turkish Wrasse *Thalassoma pavo*   p. 226

Adult male:
diagonal red and blue stripe across front of
body.
20 cm.

Turkish Wrasse *Thalassoma pavo*   p. 226

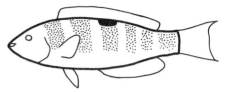

Adult female and young:
black spot below mid-point of dorsal fin;
light vertical bands.
20 cm.

Cleaver Wrasse *Xyrichthys novacula*
                                          p. 226

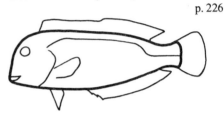

Steep forehead;
body laterally flattened.
20 cm.

Green Wrasse   p. 218
Labrus turdus L.
= *Labrus festivus* Risso
= *Labrus viridis* L.

Mediterranean.

Body long, the greatest depth of the body
is less than the length between the snout and
the hind corner of the gill cover; *lips* fleshy;
teeth conical in a single row, 3 plates of
pharyngeal teeth; *front gill cover* not
toothed; *scales* 41-49 along lateral line; *fins*
D XVII-XIX, 10-14; A III, 9-12.
The colour is very variable and has led to
much confusion. Sometimes completely
green with emerald tints or yellow-green
with a white longitudinal band running
from eye to tail, or mottled wine-red or

orange without any green but with white stripes and spots.
Up to 45 cm.

This is a coastal species living near rocks and sea grass (*Posidonia*). The larger specimens live deeper than 15 m. Has the habit of lying on its side on the bottom and in this condition it is easy to approach. Most specimens shorter than 27 cm. are female; above 38 cm. they are chiefly male. Spawns in winter and spring but little else is known. The food is made up of bottom dwelling animals of all kinds.

Brown Wrasse p. 218 pl. 63
Labrus merula (L.)

Mediterranean and Eastern Atlantic from Portugal to the Azores.

Body oval, the maximum height of body is greater than the distance from the snout to rear corner of the gill cover; *mouth* very small; *lips* fleshy; *teeth* are conical, arranged in a single series in each jaw, no pharyngeal teeth; *front gill cover* smooth; *scales* distinct but rather small, there being 40-48 along lateral line; *lateral line* runs parallel to the back; *fins* tail fin often has a distinct notch in the centre. D XVII-XIX, 11-14; A III, 8-12.
Less variable in colour than other members of the genus, usually olive or bluish-grey sometimes with a blue spot in the centre of each scale. Dorsal, anal and tail fins have a blue rim. The pectoral fin often has orange-brown rays. In the breeding season all the blue tints are accentuated.
Up to 45 cm.

A coastal species living in water of moderate depth near rocky reefs and sea grass (*Posidonia*) meadows. They are not swift swimmers and do not stray far from some refuge amongst rocks or in clumps of sea grass. The breeding season is in late winter or early spring, the male behaving very pugnaciously towards possible rivals. The eggs are generally laid on sea grass fronds. They feed on echinoids, molluscs, crustacea and polychaetes.

Ballan Wrasse p. 218
Labrus bergylta Ascanius pl. 60, 61

North East Atlantic from Canary Islands to North Scotland, rarely in the Western Basin of the Mediterranean.

Body rather long but massive with a large head; *lips* thick and fleshy; *teeth* strong and conical of moderate size, a single row in each jaw; *front gill cover* smooth; *scales* rather large, 41-47 along lateral line; *lateral line* runs parallel to the back; *fins* D XVIII-XXI, 9-13; A III, 8-12.
The colour depends on age, maturity, 'emotional state', reproductive condition and habitat and may be greenish brown, sometimes reddish and occasionally conspicuously mottled with white spots. Each scale has a lighter centre with a darker hind rim which makes them appear very prominent. The fins tend to be the same as the body colour but with white spots; some specimens have a light lateral stripe. The young fish may be emerald green.
Up to 40 cm., rarely up to 60 cm.

Most often found near steep underwater rocky cliffs and reefs down to some 20 m. They rarely venture far from crevices in the rocks. The young are occasionally found in rock pools.
Judging by its stomach contents and teeth this wrasse feeds on encrusting molluscs etc. which it bites and sucks from the rocks. Spawning is in early summer. The young fishes are common inshore in early autumn. Sudden periods of cold may kill large numbers of Ballan Wrasse.

Axillary Wrasse pl. 62
Crenilabrus mediterraneus (L.) p. 218

Mediterranean.

Mouth small; *jaws* upper slightly longer than lower; *lips* moderate; *teeth* two large conical teeth jutting forward in the front of each jaw, those of the lower jaw being somewhat smaller than the upper. On each side of the upper jaw there is a small conical tooth (sometimes absent), and in the sides of the lower jaw there are 3-4 teeth; *front gill cover* toothed; *scales* 30-35 along the lateral line; *lateral line* follows parallel to the back; *fins* D XV-XVIII, 9-11; A III, 8-12.

It has 2 conspicuous dark spots, one at the base of the pectoral, the other on the upper part of the tail stalk. In the male the pectoral spot is blue with a yellow margin, in the female it is dark brown. The normal background colour of the male is orange-brown, in the female it is slightly less red. The urinogenital papilla in the female is prominent and black; in the male it is white. In the breeding season the male is a violet-red with iridescent blue spots and stripes.
Up to 15 cm.

Lives around the deeper meadows of sea grasses (*Posidonia* and *Zostera*). Little else is known about it save that it probably breeds in late spring and summer.
The young live near the coast but move off into deeper water in winter and as they get older.

Black-tailed Wrasse

Crenilabrus melanocercus (Risso)
p. 218
pl. 183

Mediterranean.

Body oval; *eyes* large equalling in diameter the length of the snout; *mouth* small, *teeth*, 6-14 in the upper jaw, 12-18 in the lower; *front gill cover* deeply toothed on rear edge; *scales* rather large; about 30-38 along the lateral line, three rows of scales on the cheek below the eye; *lateral line* follows the contour of the back; *fins* D XV-XVII, 6-10; A III, 8-11.
The young and females are pinkish-brown above, pinkish-yellow beneath. The males are dull blue above shading to yellowish beneath. The characteristic marking of this fish is the broad dark vertical band cover-

ing the rear 2/3 of the tail fin which is edged with white.
Up to 14 cm.

Over sea grass (*Posidonia*) not deeper than 10 m. They have been observed to clean other fishes by removing parasites, fungal growths etc. Breed in spring and summer.

Corkwing Wrasse

Crenilabrus melops (L.)
pl. 64, 65, 66
p. 218

Mediterranean; East Atlantic south to Azores, north to Faroes; English Channel.

Body rather deep; *front gill cover* toothed; *scales* large (31-37 along lateral line); 5-6 rows of scales on the cheeks below the eye; *teeth* in each jaw pointed, slightly curved and arranged in a single row; *fins* D XIV-XVII, 8-11; A III, 8-11.
The colouration varies with sex, maturity and season. There is always a dark spot on the tail stalk either on, or slightly below the lateral line, but this spot may be partly obscured in the darker males. The sexes differ in colour. The females are a dull mottled brown. The males are often a dark red-brown with iridescent blue bars on the gill covers.
Up to 6 cm., occasionally up to 25 cm.
The Corkwing is the most commonly seen wrasse of nothern European waters. It is frequent in shore pools but is more often found amongst weed below low tide mark. In the winter it moves into slightly deeper water. The food is chiefly small crustacea and molluscs. The nest is built out of algae and spawning is in the spring and summer. Fishes of 5-8 cm. are about one year old. The three-year-olds are about 15-17 cm. and are mature.

Ocellated Wrasse

Crenilabrus ocellatus (Forskal)
pl. 68, 69
p. 219
Mediterranean.

Body oval; *mouth* small with prominent

lips and the upper jaw slightly longer than the lower; *lateral line* follows the contour of the back to the level of the last dorsal ray where there is a sharp downward step and from thence the lateral line is straight to the tail fin; *scales* rather large there being 30-34 along the lateral line, there are 3 rows of scales on the cheek below the eye; *front gill cover* finely toothed; *fins* D XIII-XV, 8-11; A III, 8-11.

There is always a spot on the gill cover and another dark spot at the base of the tail stalk. The colours are exceedingly variable; the ground colour is usually greenish or brownish. In the adults the spot on the gill cover is rimmed with red and (probably in the male) may be an iridescent blue.
Up to 13 cm.

Lives at moderate depths near rocks and sand. Probably breeds in late spring and early summer building a nest of seaweed. The eggs are adhesive. These fish have been observed to clean parasites from the bodies of other fishes.

Painted Wrasse pl. 71, 72, 73
Crenilabrus tinca (L.) p. 219

Mediterranean.

Body oval and flattened; *eye* small; snout long, 3-4 times diameter of eye; *profile* long, concave above the eye with a slight convex hump on the snout; *lips* very fleshy and protruberant; *teeth* small and almost hidden by soft tissue, there are 8-24 in the upper jaw and 10 or more in the lower; *front gill cover* finely toothed; *scales* large with 33-38 scales along the lateral line and 5 series of scales on cheek below the eye; *fins*, soft rays of the dorsal and anal fins are much longer than the spiny rays. D XIV-XVI, 10-12; A III, 8-12.

Very variable in colour. Usually the ground colour is yellow-olive with red and blue stippling arranged in longitudinal bands on the flanks. There is a dark spot at the base of the tail fin and another at the base of the pecto-ral. A dark band runs between the eyes across the snout just above the level of the upper lip.

Males up to 33 cm., females up to 25 cm.

This species is known to be active only during the day, but most wrasse behave in this way. Lives near rocks and sea grass (*Posidonia*) meadows. Spawn in summer; the male makes a nest out of algae, the female attaches her eggs to the fronds and the male fertilizes them. They feed on small bottom dwelling animals.

Five-spotted Wrasse pl. 70
 p. 219
Crenilabrus quinquemaculatus (Bloch)

Mediterranean.

Body deep, its greatest depth 3-3 1/2 times total length; *teeth* conical, 10-12 in the upper jaw, 10-16 in lower; *front gill cover* finely toothed; *scales* very large, 30-33 along lateral line, 4 series of scales beneath the eye; *lateral line* follows contour of back until the last few rays of the dorsal fin when there is an abrupt downward step and thereafter is straight to the tail; *fins* D XIV-XV, 9; A III, 8-9.

Colour very variable; generally whitish-brown with large darker mottling or it is distinctly dark and light spotted. The male has red lips and the body is flecked with greenish-blue. There is often a black spot at the base of the tail stalk on the lateral line. During spawning the female genital papilla is bright blue.
Up to 15 cm.

Lives on rocky bottoms sometimes in water less than 1 m. deep even where there is considerable surge. Breeds in early summer. The nest is built by the male who will not tolerate the presence of the female at this time. The nest is a hemispherical mound with a cavity near the bottom. When the nest is built the female enters the cavity lays her eggs and the male fertilizes them. He then chases off the female. The exterior of the nest is covered with the flat fronds of the sea-lettuce (*Ulva*) and after the eggs are laid the whole structure is covered with gravel. The male guards the eggs.

Crenilabrus doderleini Jordan   p. 219

Mediterranean.

Body long, its length being about 4 times its maximum height; *mouth* considerably longer than eye diameter; *teeth* small and conical, about 12 in the upper jaw, 14 in the lower; *front gill cover* finely and shallowly toothed; *scales* large, 30-36 along the lateral line and 3 rows of smaller scales beneath the eye; *lateral line* follows contour of back along its entire length; *fins* D XIII-XVI, 9-11; A III, 8-10.

Reddish or chestnut brown, sometimes with white spots. There is a longitudinal dark band running from the snout to the tail stalk with a lighter band beneath. There is a dark spot on the tail stalk above the lateral line.

Up to 10 cm.

Rare, lives from about 3-12 m. near rocks and sea grass (*Posidonia*). Breeds in spring.

Grey Wrasse   p. 219 **pl. 75**
Crenilabrus cinereus (Bonnaterre)

Mediterranean.

Body oval, head as long as the maximum depth of body; *mouth* small; *teeth* 20-28 in the upper jaw of which the front ones slant forwards, there are 28 teeth in the lower jaw; *front gill cover* finely toothed; *scales* rather large, 31-35 along the lateral line; 2 rows of large scales beneath the eyes; *lateral line* smoothly curved but the hind end lies higher on the flank than the front end; *fins* D XII-XV, 9-11; A III, 8-10.

Usually grey with white spots arranged in a longitudinal band; often mottled brown or with longitudinal brown bands. There is a dark spot set low on the tail stalk below the lateral line. The mature males have golden tints on the flanks and throat and the cheeks have gold and blue stripes. The urinogenital papilla is colourless in the male but bright blue in the female.

Up to 8 cm., sometimes up to 16 cm.

Common on sandy and muddy bottoms and seems to prefer estuaries and enclosed bays where there is plenty of vegetation and detritus. Spawns in early summer, the nest being similar to that of *Crenilabrus quinquemaculatus* but is more firmly glued together. Feeds on crustacea and molluscs.

Long-snouted Wrasse   p. 219
Crenilabrus rostratus (Günther)   **pl.76**
= *Crenilabrus scina* (Forskal)

Mediterranean.

Body rather elongate; *snout* very long giving a concave profile; *mouth* small; *teeth* a single file of small sharp teeth in each jaw; *front gill cover* finely toothed on its hind edge; *scales* rather large 30-35 along the lateral line; 3-4 rows of scales beneath the eye; *lateral line* follows the contour of the back to the last dorsal ray when there is an abrupt downward step and thereafter it is straight to the tail; *fins* D XIV-XVI, 9-12; A III, 9-11.

Colour variable; reddish-brown, mottled brown and black or green. The sexes are not greatly different in colour.

Up to 12 cm.

Lives in water deeper than 15 m. near sea grass (*Posidonia*). The eggs are laid on the bottom in early summer.

Rock Cock   **pl. 65, 67**
Centrolabrus exoletus (L.)   p. 219

North East Atlantic from Biscay to North Scotland and Norway. It is not present in the southern North Sea.

Mouth small reaching half way back to eye; *teeth* conical in a single row in each jaw; *front gill cover* toothed on rear and lower

224

edge; *lateral line* 33-37 scales along lateral lines; *fins*, usually 5 anal spines. D XVIII-XX, 5-7; A IV-VI, 6-8.

Usually a warm red-brown above, yellowish silver on the flanks and silver below. Males are flecked with iridescent blue and have two iridescent blue bands passing from the eye to the angle of the mouth. On the tail there is a broad dark band; the rim of the tail is white.

Rarely up to 15 cm.

Usually amongst kelp-covered rocks beneath low-tide mark. Probably spawns in summer.

Goldsinny   p. 219 **pl. 81**
Ctenolabrus rupestris Cuv. et V.
= *Ctenolabrus suillus* (L.)

Mediterranean; East Atlantic north to Norway. English Channel.

Body rather long; *teeth*, several rows in each jaw, the front ones are curved and inclined slightly forward; *front gill cover* very finely toothed but smooth on the lower edge; *scales* 34-40 along lateral line; *lateral line* parallel to back; *fins*, the dorsal fin is of uniform height throughout its length. D XVI-XVIII, 8-10; A III, 7-10.

The adults are a red-brown colour; the young may be a fawn to dull green. There is always a dark spot on the top edge of the tail stalk.

Up to 18 cm.

The Goldsinny does not come into very shallow water being found chiefly amongst rocks. They are said to move into deeper water in winter but it is more likely that they are just more difficult to find.

They spawn in summer and feed on bottom living crustacea, molluscs etc.

Lappanella fasciata (Cocco)   p. 220

Western basin of the Mediterranean and Madeira.

Resembles *Crenilabrus rostratus* in general form but there is no step in the lateral line; it has distinct canines and there are 35-38 scales along the lateral line. *Fins* D XVI-XVII, 8-12; A III, 8-11.

Pinkish-brown above, yellowish-pink below. There is a black spot on the soft rays of the dorsal fin and one on the middle rays of the tail fin. The sexes are not obviously different.

Up to 14 cm.

Rare. Found on muddy bottoms deeper than 60 m.

Rainbow Wrasse   p. 220 **pl. 77**
Coris julis (L.)

Mediterranean and East Atlantic from Guinea to Biscay.

Body long and sinuous; *snout* long and pointed; *teeth* sharp and pointed, inclined forward in the front of the jaws; *eye* small; *scales* very small (73-80 along the lateral line); no scales beneath the eye; *fins*, first 2-3 spiny dorsal rays are elongate in the male. Dorsals and anals are long and of uniform height throughout their length. D VIII-IX, 12; A III, 11-12.

There are two distinct colour forms; the smaller female phase has a blue spot on the lower edge of the gill cover. Specimens caught from water less than about 20 m. have a dark olive-brown ground colour, those from deeper water are red or red-brown. There is a longitudinal yellow stripe running from nose to tail fin. To the diver the colouration appears the same irrespective of depth as the increased reddening of the fish compensates for the loss of red light penetrating the water. The male phase has a black and orange spot on the elongated dorsal rays. There is a zig-zag orange mark along the flanks and a lozenge shaped black mark behind the pectoral fin.

Intermediate colour forms are common and the female form sometimes turns out to be the male and vice-versa.

Males up to 25 cm., females up to 18 cm.

225

Very abundant in the Mediterranean, they live around weed-covered rocks and over sea grass (*Posidonia*) from near the surface to at least 120 m. In winter they move off into deeper water.

Like most (perhaps all) wrasse they are active only during the daytime. The young of this species has been observed to clean parasites from other fishes.

In the aquarium they dig themselves into the sand at night; but as far as we know they have never been seen to do this in nature and where they spend the night remains a mystery.

This is an hermaphrodite species and females often become males. Breed in early summer and the eggs float.

They feed on small animals.

Turkish Wrasse; Rainbow Wrasse; Peacock Wrasse p. 220 **pl. 78, 79, 80**
Thalassoma pavo L.

Mediterranean and North East Atlantic from Biscay to Guinea.

Body somewhat elongate; *mouth* small; *teeth* 1 row of small sharp teeth in each jaw; *scales* large, 26-31 along the lateral line; *lateral line* follows the contour of the back until the 10th-11th soft dorsal ray when there is an abrupt downward step and from thence to the tail fin it is straight; *fins* D VIII, 12-13; A III, 10-12.

Two main colour forms. The most common has a black spot on the back beneath the centre of the dorsal fin; it is generally moss-green in colour with lighter vertical bands. A less common form (probably the male) has a diagonal red and blue band running from the front of the dorsal fin to the pelvic. The flanks are green and the head is dark red with blue veins. An uncommon intermediate form has neither a dark dorsal spot nor diagonal blue and red lines.

Up to 20 cm.

Often very common living around weed covered rocks to about 20 m.

Spawns in summer and the eggs float. The sex of an individual fish may change in this species, the change from female to male may be induced by warmer weather.

Cleaver Wrasse p. 220 **pl. 74**
Xyrichthys novacula (L.)

Mediterranean and East Atlantic from Biscay to Guinea.

Body deep and laterally flattened; *profile*, almost vertical forehead; *eye* is small set high in the head; *mouth* set very low; *teeth* in the front of the jaw are very sharp and are slanted forward; there are grinding teeth in the back of the jaw; *scales* large 24-29 along the lateral line; *lateral line* follows the contour of the back to the last dorsal rays where there is a pronounced downward step and from thence continues straight to the tail; *fins* D IX-X, 11-12; A III, 11-13.

The females are predominantly red and the males chiefly grey-green. They may be very pale in colour over sand.

Up to 20 cm., sometimes up to 30 cm.

Lives on sand and mud bottoms at around 5-10 m. They may take refuge in holes and are also able to bury themselves very rapidly in the sand. Spawn in summer and move into deeper water during the winter.

Family:

SCARIDAE

In our area *Euscarus cretensis* is the only representative of this family, but in tropical waters there are numerous species many of which are common. The teeth in each jaw are fused into a parrot-like beak and are used to scrape the hard surface of rocks and coral. On tropical reefs the parrot fishes are a significant force in grinding the coral to sand.

Parrot Fish   pl. 84 pl. 184
Euscarus cretensis (L.)
= *Sparisoma cretense* Jordan & Gunn

Mediterranean and Eastern Atlantic from central Portugal to the Canary Islands.

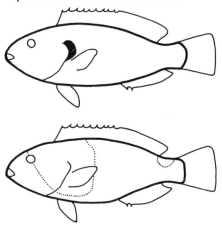

Teeth fused into parrot-like beak; very large scales.
Two colour forms:
1, grey with black spot behind gill cover
2, red with yellow spot at top of tail stalk.
   Grey patch on shoulder.
50 cm.

Body oval; *snout* blunt; *teeth* are fused together in each jaw giving a parrot-like beak; *scales* are very large only 23-25 along the lateral line. A single row of scales on the cheek below the eye. D X, 10; A II, 10.
There are two colour forms. One (probably the male) is grey with a large black blotch behind the gill cover. The other (probably the female) has a red ground-colour but with a large grey patch on the shoulder with yellow blotches on the gill cover in front of the pectorals and another on the top of the tail stalk. The eye is also yellow.
Up to 50 cm., commonly 30 cm.

Usually seen in groups of three or four which may contain individuals of both colour forms. They frequent rugged rocky areas especially underwater cliffs. They are not strong swimmers and scull themselves along with their pectoral fins. Parrot Fish scrape the encrusting algae from the rocks with their strong beak and are also said to scrape the algal growth from sea grass (*Posidonia*) fronds.

Family:

TRACHINIDAE

Elongate bottom living fish which have venom glands at the base of the spiny rays in their first dorsal fin and at the base of a strong spine on their gill cover. The scales are arranged in diagonal rows.
There are four species in our area. They all spend most of the day buried in the sand with only their head and first dorsal fin exposed. When alarmed the first dorsal fin is raised and can cause a wound so painful that it can be incapacitating and is thus dangerous. Most cases of permanent injury, however, are the result of the wound becoming septic. The best treatment known at the present is to hold the affected part in as hot water as can possibly be borne for at least half an hour. Afterwards the wound should be cleaned, disinfected and covered in the normal way. This hot water treatment is effective because the venom from the Weever is destroyed by heat.
The Weevers feed on small fishes of all kinds and it is thought that they may come out more into the open at night.

Spotted Weever *Trachinus araneus*   **pl. 83**

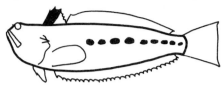

7-11 dark spots along flanks;
7 rays in 1st dorsal fin.
25 cm.

Mediterranean.

Body long and laterally flattened; *mouth* almost vertical in the head; *spines*, there is a pair of small spines on the head just in front of the eyes; *fins* there are 7 poisonous spine rays on the 1st dorsal. 1D VII 2D 26-29; A II, 29-31.
Brown or yellow-brown above, paler below with 7-11 darker spots in a longitudinal series along the flanks. Front half of 1st dorsal fin black.
Up to 25 cm., exceptionally up to 40 cm.

A shallow-living species. They are generally found half buried in sandy bottoms especially where there are rocks or sea grass nearby.

Greater Weever *Trachinus draco*   **pl. 186**

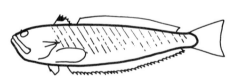

Oblique pattern on flanks;
5 rays in 1st dorsal fin.
40 cm.

Mediterranean and Eastern Atlantic from Morocco to Norway.

Body elongate, shallow and laterally flat-

tened; *mouth* terminal and oblique; *spines*, a small spine on head in front of each eye. A solitary venomous spine on each gill cover; *fins*, the 1st dorsal fin usually has 5 or 6 strong venomous spines. 1D V-VII; 2D 29-32; A II, 28-34.
Greyish yellow above, paler below with characteristic narrow diagonal streaks on the flanks. The dorsal fin has a large black spot.
Up to 40 cm.

Lives on sandy bottoms down to 100 m. or more but may occasionally come into water more shallow than 10 m.
Breeds in late spring and summer. The eggs are free-floating.

Streaked Weever *Trachinus radiatus* **pl. 185**

Irregular dark spots and circles;
plaques with radiating grooves on head.
25 cm.

Mediterranean and Eastern Atlantic from Morocco to Senegal.

Body long and laterally flattened; *head*, on the head behind the eyes are 3 large bony plaques with radiating ridges; *mouth* terminal and nearly horizontal; *spines*, there are 2 or 3 small spines on the head just in front of the eyes and a large venomous spine on each gill cover; *fins*, the 1st dorsal fin has 6 venomous spines. 1D VI; 2D 24-27; A I, 26-29.

Brown above and yellow-brown below. The area beneath the head is whitish. There is a characteristic pattern of dark rings and spots. The membranes between the first 4 dorsal spines are black.
Frequents sandy bottoms. Probably lives

228

down to about 200 m. and is generally a deeper-living fish than the other Weevers.

Lesser Weever *Trachinus vipera*   **pl. 82**

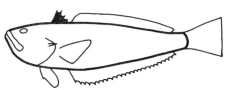

Markings faint or absent;
dorsal fin black;
no small spines on head.
14 cm.

Mediterranean and Eastern Atlantic from Senegal to Scotland.

Body long and laterally flattened but the body is rather deeper than the other weevers; *mouth* terminal and oblique; *spines* there are no spines on the head. There is a long venemous spine on the gill cover; *fins*, the 1st dorsal fin normally has 6 spines. 1D V-VII; 2D 21-24; A I, 24-26.

Yellowish brown above, lighter below. No conspicuous pattern but there are small dark spots on the head and back. The dorsal fin is entirely black.

Up to 14 cm.

Found on sandy bottoms from about 1 m. down to 50 m. During the daytime they lie with their eyes and dorsal fins exposed and can cause excruciating pain to a bather if he treads on one. A more serious menace is to inshore trawlermen, especially shrimpers, who get stung when sorting their catch. There is a report that this fish has actually attacked a swimmer.

Family:

URANOSCOPIDAE

Body has a solid conical form with a flat top to the head. The eyes are set on the upper surface of the head. There is a strong spine behind the gill cover and an electric organ behind each eye.

Star Gazer *Uranoscopus scaber*   **pl. 85, 86**

Stiff conical body;
flat top to head;
mouth nearly vertical.
25 cm.

Mediterranean; Eastern Atlantic from Spain to Senegal.

Body elongated and rounded; *head* laterally flattened with rough bony plates; *eye* very small and set on the top of the head; *mouth* large and almost vertical. On the tip of the lower jaw is a small appendage; *spines*, there is a sharp venemous spine behind the gill cover; *fins* two dorsal fins. The 1st has 4 spiny rays. 1D IV; 2D 13-14; A I, 13-14.

Dull brown in colour, darker above, lighter beneath.

Up to 25 cm., rarely 30 cm.

The Star Gazer lies almost buried in the sand with only its mouth and eye visible. The small process on the lower jaw can be made to vibrate, luring small fishes within reach of its jaws.

Behind the eye on each side is an oval electric organ. Out of water the Star Gazer is able to give an unpleasantly powerful electric shock. Underwater this shock would not be so strong. Breeds in spring and summer. The eggs are free floating.

Sub-order: TRICHIUROIDEI

Oceanic fishes with long or very long laterally flattened bodies and strong well-developed jaws. Many are bathypelagic.

Family:

TRICHIURIDAE

Very elongated and laterally flattened body. A single dorsal fin runs the entire length of the back. The tail fin much reduced.

Scabbard Fish
Lepidopus caudatus (Euphrasen)

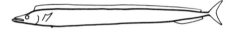

Long flattened body;
silvery.
2 m.

Mediterranean; East Atlantic north to Britain; western English Channel.

Body long and laterally flattened; *head* long, about 1/7 total length; *profile* gently slopes from snout to behind the eye then rises more steeply to the 1st dorsal fin; *snout* pointed; *jaws*, lower jaw longer than upper; *teeth* sharp and pointed with some much larger teeth situated towards the front of the jaws, 4-6 in the upper jaw, 1-2 in the lower; *scales* this fish is completely without scales;

fins, 1 dorsal fin commencing behind the head and ending just in front of the tail fin. Anal fin short, equal to the length of the head and situated at the rear of the body. There are anal fin spines but these are more ·r less embedded in the body tissue. Pelvic fins are present as small scales behind the pectoral fins. Tail fin deeply forked. D C-CV; A XXXI-XXXVI, 21-25.
This fish is completely silver though there may be a golden stripe along the flanks.
More than 2 m.

A mid-water fish which is found from the surface down to 400 m. Usually in shoals and frequently caught over sand. They are carnivores feeding on small fish and crustacea. Spawning occurs off the North African coast from the end of winter to spring. This fish is particularly common off the Portugese coast where it is of commercial importance.

Family:

SCOMBRIDAE

Strong and active swimmers that may undertake long migrations. The **Mackerels** are chiefly known as free swimming fishes living in large shoals near the surface where they feed on smaller fishes. They are abundant and good to eat but like their close relatives the Bonitos and Tunnies they should always be cooked soon after capture for their flesh seems particularly prone to a type of bacterial decay that makes it poisonous. The scales are very small but cover the body. Between the head and about the centre of the first dorsal fin there is an irregular band where the scales are larger (the corselet) but this is not well developed in the Mackerel (*Scomber scombrus*). The two dorsal fins are well separated and behind both the anal and 2nd dorsal fins there is a series of finlets. The tail fin is divided into two distinct lobes.

Atlantic Mackerel *Scomber scombrus*

'Mackerel-stripes' along entire back;
widely spaced dorsal fins;
no corselet.
50 cm.

Chub Mackerel *Scomber colias*   p. 232

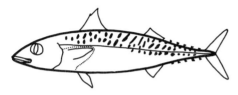

Yellow stripe down body;
corselet;
widely spaced dorsal fins.
30 cm.

Mackerel   **pl. 87**
Scomber scombrus L.

Mediterranean and north Atlantic. There is a closely related species in the Pacific.

Body rather long and streamlined; *eye* large with vertical transparent eyelids; *keels* none along the mid-line of the tail stalk but there are two small keels on the base of each lobe of the tail fin. No keel between the two dorsal fins; *teeth* small; *lateral line* gently curved; *scales* no corselet; *fins* the two dorsal fins are well spaced, there is a series of 5 pinules behind the 2nd dorsal and the anal fins. Tail fin in 2 lobes. D X-XIII; D 11-13 + 5 finlets; A I, 11-13 + 5 finlets.
Iridescent green and blue above, whitish silver below. There are irregular zebra-like

dark stripes on the back and upper flanks. Up to 50 cm. sometimes much more.

In the late spring through the summer to early autumn great shoals of mackerel often come close inshore. In the winter they move into deeper water offshore usually to the edge of the continental slope where they keep close to the bottom and probably eat very little.

In the warmer months their food is very variable consisting of swimming crustacea, fish fry, young herrings etc. They will snap at almost anything and a wide variety of objects such as feathers and silver paper can be used to hook them.

Spawning is in spring and summer and the eggs float. Because of this prolonged breeding season members of one year class are very variable in size but at one year the average length is about 24 cm. They mature in their second year when they are about 30 cm. long. It is thought that some mackerel may live for 20 years.

Very swift swimmers they can easily outdistance a commercial trawl although many are caught in this way. In the aquarium they are in constant motion for they rely for their oxygen supply upon the water current set up through their gills by their passage through the water.

There is considerable mystery about their camouflage. Sometimes, especially when they are being chivvied along by a trawl they are quite conspicuous, especially when seen from above. In the open sea they are virtually invisible. Not surprisingly we have no photographs of mackerel swimming free in the ocean. The specimens in pl. 87 are in an aquarium and show the curious and unnatural massive growth of the lower jaw frequently found in specimens kept in the aquarium.

Very good to eat but should be fresh.

Chub Mackerel; Spanish Mackerel   p. 231
Scomber colias Gmelin
= *Scomber japonicus* Houttuyn

Atlantic and Mediterranean.

Resembles the Mackerel *Scomber scombrus* in general body form but differs in having no swim bladder, 7-9 spiny dorsal rays and a corselet around the head and pectoral base.
The flanks and back are marked with spots rather than stripes and there is a golden band running from the gill cover to the tail.
Up to 30 cm., occasionally 60 cm.

Lives in somewhat deeper water than the mackerel. At night they can be attracted by surface lights perhaps because the lights also attract swarms of pelagic crustacea on which they feed.
A shoaling pelagic species they can undertake long migrations. They chiefly feed on small fishes and swimming crustacea. They come into breeding condition in summer but the exact time depends on the water temperature. In the Mediterranean the breeding grounds are in the Sea of Marmora. They probably do not breed in the northern part of their range.
An important food fish in the Mediterranean but too spasmodic in northern Europe to have any commercial importance.

Family:

SCOMBEROMORIDAE

Resemble the Scombridae except that the two dorsal fins are set very close together. The corselet of scales is well developed. The tail is moon shaped not divided into distinct lobes.

Atlantic Bonito; Pelamid
Sarda sarda (Bloch)

Longitudinal stripes on upper flanks;
wavy lateral line;
adjacent dorsal fins.
80 cm.

Mediterranean and tropical Atlantic, rarely north to Sweden.

Body elongate; *jaws* long reaching to the hind edge of eye. The upper jaw being slightly the more prominent; *teeth* robust;

keels there is a keel on each side of the tail stalk flanked on each side by small longitudinal ridges converging towards the tail; *scales* well developed corselet present; *lateral line* wavy; *fins*, the 2 dorsal fins are adjacent. 1D XXI-XXIV; 2D 12-16 + 7-10 finlets; A 11-15 + 6-8 finlets.
Azure or ultramarine on the back, silvery white below. The upper part of the flanks are marked with longitudinal or oblique stripes.
Up to 60 cm. in northern waters, 80 cm. in the south.

A strongly migratory fish swimming in shoals usually near the surface and some miles offshore. They are only caught from waters between 15° and 24°C. Do not spawn off northern Europe, but in the Mediterranean they spawn in the winter between November and June.

232

Family:

THUNNIDAE

The **Tunnies** are one of the most important food fishes of the world and are found in all warm seas. These fishes are not strictly speaking cold blooded for on capture their blood temperature is several degrees above the surrounding water.

Members of this family have rather deep bodies that only have scales along the lateral line and in an irregular band (the corselet) around the body between the head and rear portion of the 1st dorsal fin. The two dorsal fins are separated by a short space. The tail fin is moon-shaped, not divided into two distinct lobes.

Like the Mackerels (Scombridae) the flesh should be cooked soon after capture for it is prone to a type of bacterial decay that renders it poisonous.

Tunny *Thunnus thynnus*

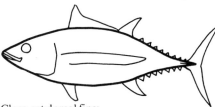

Dorsal fins set close together;
no markings on body;
pectoral fin short.
3 m.

Albacore *Thunnus alalunga*   p. 234

Close-set dorsal fins;
long pectoral fin.
1 m.

Little Tunny *Euthynnus alletteratus*   p. 234

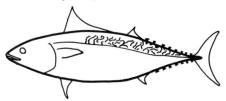

Dorsal fins close together;
Mackerel-like marking above lateral line
and behind pectoral fin.
80 cm.

Skipjack Tuna
Euthynnus (Katsuwonus) pelamis   p. 235

Dorsal fins close together;
longitudinal stripes behind pectoral fin and
below lateral line.
65 cm.

Frigate Mackerel *Auxis thazard*   p. 235

Widely spaced dorsal fins;
Mackerel-like markings behind 1st. dorsal
fin and above lateral line.
60 cm.

Tunny; Blue-Fin Tuna   **pl. 189**
Thunnus thynnus (L.)
Mediterranean and both sides of the Atlantic from Norway to South Africa. A similar species inhabits the Pacific and Indian Oceans.

233

Body rather deeper than the mackerels; *jaws* rather small and extend back to the level of the eye; *teeth* small, sharp and conical arranged in a single file in each jaw; *keels* there is a well-developed keel on each side of the tail stalk; *scales* small on most of the body but large in the region of the corselet; *fins* the 2 dorsal fins set closer together than the eyes's diameter. Pectoral fin short and does not reach further back than the 12th dorsal spine. 1D XIII-XV; 2D I, B-15 + 8-10 finlets; A 11-15 + 8-9 finlets.
Blue-black above, silvery white below. The fins are dark except for the anal fins and finlets which are dull yellow. Sometimes the pelvic fins are yellow as well.
Up to 3 m.

The Tunny together with its near relative in the Pacific is of great commercial importance. They embark on long migrations and the presence of large shoals in particular places are predictable. They are known to cross the Atlantic and in the summer some migrate from the Mediterranean to Norway and the North Sea. They do not inhabit water colder than about 10-14°C and are not found in northern European waters during the winter. In the summer they school near the surface and may come close inshore but in winter they live deeper, sometimes to 180 m.
Tuna breed in the warmer parts of their range and the pelagic eggs are laid in summer. A fish hatched in June may weight 0.5 kilos by September and an old fish 18 years old may weigh 200 kilos. They feed on small fish and squids and the young also take various swimming and bottom-living crustacea.
The Tuna is of enormous commercial importance and is also a favourite game fish. They are trapped in nets set in the path of their migrations and are also caught on lines. Tuna meat is generally canned in_oil and is eaten all over the world.

Albacore; Long-finned Tunny   p. 233
Thunnus alalunga (Bonnaterre)

Mediterranean and all warm oceans, rarely further north than Biscay.

Resembles the Tunny except that the pectoral fin is very long extending back at least as far as the rear edge of the 2nd dorsal fin.
Bluish-black or bronze-brown above, whitish below with an iridescent blue band running along the flanks. The margin of the tail fin is light and may show up well underwater. The finlets are yellow.
Up to 100 cm.

Like the Tuna these are a social and migratory species but keep to water warmer than 14°-18°C. During their summer migrations they do not normally go further north than Biscay but there they are common. They do not come close inshore and can be caught from as deep as 200 m. off the Florida coast. Tuna have a body temperature distinctly warmer than the surrounding water. Healthy individuals of this species may have a body temperature 9°-18.7°C warmer than their environment.
An extremely important commercial species in southern European waters. They are caught both on lines and are trapped in nets.

Little Tunny; Bonito   p. 233
Euthynnus alletteratus (Rafinesque)

Mediterranean and the warm waters of the Atlantic as far north as Portugal.

Body resembles the other Tunnies in general shape; *scales*, no scales except in the region of the corselet and along the lateral line; *keels* one distinct keel on each side of the tail stalk; *fins*, the two dorsal fins are separated by less than the eye's diameter. The pectoral fin is short and the pelvic fins are joined at the base by the so-called interpelvic processes. 1D XIII-XVI; 2D II-14 + 8 finlets; A 12-14 + 7 finlets.
Ground colour is blue-back above, whitish below. There is a panel of mackerel-like lines and spots on the back behind the tip of the pectoral fin. There is often a pattern of

about 6 dark spots behind and below the pectoral fin.
Up to 80 cm.

A gregarious warm water species chiefly feeding on pelagic fishes.

Skipjack Tuna; Ocean Bonito   p. 233
Euthynnus pelamis (L.)   **pl. 187, 188**
= *Katsuwonus pelamis* Kishinouye

Mediterranean and warm areas of all oceans. In the north-east Atlantic north to west coast of England.

Body oval and rather deep of the typical Tunny shape; *scales* there are no scales except in the region of the corselet; *fins* the two dorsal fins are set close together. The pectoral is short not extending beyond the 9th dorsal ray. 1D XV-XI; 2D 11-16 + 8 finlets; A 11-16 + 7 finlets.
Blue-black above, whitish below with 4 to 6 dark longitudinal stripes below the lateral line and behind the pectoral fin. Occasionally there is a series of diffuse vertical bands along the flanks.
Up to 65 cm., occasionally 90 cm.

A free swimming and gregarious species, normally prefers deep ocean water from 18°-20°C but occasionally come close inshore generally in the summer. Do not spawn in the northern part of their range and their spawning habits in the Mediterranean are not well known. In the American Atlantic and Pacific they spawn all the year round. In the Mediterranean, West Atlantic and Pacific they are important commercial and sporting fish.

Frigate Mackerel; Plain Bonito   p. 233
Auxis thazard (Lacépède)   **pl. 88**

Mediterranean and warm parts of all oceans. In the north-east Atlantic north to Biscay.

Body slightly longer than the other Tunnies; *eye* rather small and set forward in the head, the head being between 5 and 6 times the diameter of the eye; *keel* there is a well developed keel along the mid-line of the tail stalk; *scales*, there are no scales on the body except for a corselet of large scales that extend in two lobes, one along the back, the other along the lateral line to the level of the second dorsal fin; *fins*, the 2 dorsal fins are widely spaced and rather small. Between the two pelvic fins is a large triangular flap of skin characteristic of the genus. 1D X-XI; 2D 11-12 + 8-9 finlets; A 12-15 + 7-8 finlets.
Blue or blue-green above, silvery beneath. Above the lateral line and behind the corselet is a conspicuous mackerel-like panel of darker spots and lines.
Up to 60 cm., occasionally 100 cm.

A schooling pelagic species that sometimes comes close inshore during the summer. Not sufficiently common to have any commercial importance.

Family:

ISTIOPHORIDAE

The **Spearfish** and **Sailfish** have the bones of the upper jaw elongated into a long cylindrical spear whilst in the swordfish (*Xiphiidae*) it is flattened. There are two keels on each side of the tail stalk.

Marlin; Spear Fish
Tetrapturus belone Rafinesque

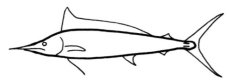

Sword cylindrical;
front part of long 1st dorsal fin angled;
uniform colour.
2 m.

Mediterranean.

Body slender, gradually tapering towards the tail and with 2 keels on either side of the tail stalk; *jaws*, upper jaw elongated to produce a round 'sword-like' structure; lower jaw slightly elongated; *profile* straight or only slightly concave; *scales* small and embedded in the skin; *fins*, 2 dorsal fins, 1st very long with the first 4 rays elongated; 2nd dorsal small and just separated from the 1st. 2 anal fins; pelvic fins very slender and consist of 1 spine and 4 rays fused together. Tail fin very slender; *vent* situated a considerable distance in front of the anal fin. 1D 39-46; 2D 6; 1A 11-15; 2A 6-7.
Back dark grey or dark blue, lighter on the sides and white underneath.
Up to 2 m.

These fish are strong swimmers and are found in mid-water. They are predatory carnivores which feed on shoals of small fish which they pursue. They are popular game fish.

White Marlin
Tetrapturus albidus Poey

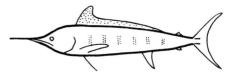

Sword cylindrical;
front part of long 1st dorsal fin rounded;
dark spots on dorsal fin and lilac bands on body.
2.5 m.

These fish are similar to the Marlin (*Tetrapturus belone*) but are also found in the East Atlantic north to Britain as well as in the Mediterranean.
They may be distinguished by the more rounded front part of the 1st dorsal fin; more concave profile and by the vent which is situated close to the anal fin. The 2nd dorsal and 2nd anal fins are very small. 1D 38-43; 2D 5-6; 1A 13-19; 2A 5-6.
Back greyish or bluish in colour, lighter on sides whitish underneath. The 1st dorsal fin has black spots and there may also be lilac coloured vertical stripes along the body.
The habits of this fish are similar to those of *Tetrapturus belone*.

Family:

XIPHIIDAE

The upper jaw of the **Sword Fish** is elongated into a long flattened sword-like weapon. There is a well-developed lateral keel on each side of the tail stalk. In adults there are two well separated dorsal fins. The first is triangular, the second is very small and resembles a finlet.

236

Sword Fish   pl. 190
Xiphias gladius L.

Sword flattened;
1st dorsal fin short.
3.3 m.

Mediterranean, throughout all tropical and
temperate oceans; sometimes English Channel; North Sea; West Baltic.

Body fairly deep and tapers gradually towards tail; tail stalk has one keel on either
side; *profile* rather concave; *jaws*, upper jaw
greatly elongated to form the characteristic
flattened 'sword' that gives this fish its
name. Lower jaw slightly elongate; *fins*
adult fish have 2 dorsal fins, the 1st is short
and high and the 2nd small. There are two
anal fins the 2nd opposite and similar to

the 2nd dorsal fin. Young fish have 1 dorsal
and 1 anal fin both of which gradually separate as the fish mature. Tail fin lobes slender.
There are no pelvic fins.
Adults 1D III-IV, 16; 2D 3-4; 1A II-III, 9;
2A 3-5.
Young D III-IV, 40; A II-III, 15.
Back uniform dark grey, dark blue or
brownish red, lighter on sides with bronze
reflections. Undersurface whitish-grey.
Up to 3.3 m., occasionally to 4.9 m.

These fish are solitary and may be found in
open water from the surface down to below
600 m. They are frequently encountered
swimming near the surface with just their
dorsal and tail fins breaking the surface.
They are voracious, fast swimming fish
which pursue and feed on shoals of fishes
such as herring and sardines. It is possible
that they catch some fish by flailing the water and fish with their swords and then eat-
ing the dead and injured fish. They breed
during the summer in water above 24°C.
They are edible and popular game fish.

Family:

LUVARIDAE

This family only has one genus containing the single species *Luvarus imperialis*.

Luvar
Luvarus imperialis Rafinesque-Schmaltz

Oval flattened body;
very steep forehead;
pink fins, black area on back.
2 m.

Mediterranean; Atlantic and all oceans of
the world.

Body very large, deep and laterally flattened
with one keel either side of the tail stalk;
profile very steep and 'hump-like'; *mouth*
very small with no teeth; *eye* rather small
and at the same height as the mouth; *fins* 1
dorsal fin and 1 anal fin which are similar
and opposite each other. Tail fin high, pelvic fins very small. D 13-14; A 13-14.
Back dark metallic bluish or black, the sides
are silvery pink or silver and the undersurface is silver. The front of the snout, the

pectoral, anal and tail fins are pink. The front of the dorsal fin is also pink and the rest blackish.
Up to 2 m.

These are solitary oceanic fish which are only very occasionally seen, usually either when stranded or when very young. Very little is known about them except that they feed on planktonic animals, and possibly small fish, molluscs and cephalopods.

Sub-order: STROMATEOIDEI

Oceanic fishes, sometimes living in very deep water. They have soft bones and fragile fins. The body is covered with fine scales that are easily detached. The snout is very blunt in outline. These fishes are distinguished from all others by the pouch on each side of the oesophagus containing many rows of soft tooth-like papillae.

Family:

CENTROLOPHIDAE

No teeth on the roof of the mouth. The oesophogeal papillae are irregular in shape and are arranged in 10-20 rows. There is a single dorsal fin.

Black Fish
Centrolophus niger (Gmelin)
= *Centrolophus pompilus* (Risso)

Black with silver highlights;
single dorsal fin.
86 cm.

Mediterranean; East Atlantic north to Iceland, south to Madeira and Azores; North sea; western Channel.

Body oval with a long slender tail stalk; *profile* curved; *snout* rounded; *jaws* quite short only just extending behind the front edge of the eye; *front gill cover* very finely toothed; *scales* very small; *fins*, 1 dorsal fin starting just behind the pectoral fin and terminating at the beginning of the tail stalk. Tail fin large and forked. D X, 28-31; A III, 21-22.

Back very dark grey or black, lighter on lower sides and under surface. The very small scales all have a dark edge and sometimes have silver patches which give the fish an overall dark appearance with silver highlights.
Up to 86 cm.

An oceanic solitary species which wanders throughout the oceans. Very little is known about them except that they feed on fish, cephalopods and crustacea. They are known to breed in the Mediterranean during late autumn.

Family:

CALLIONYMIDAE

Despite the bizarre appearance of the courting male the diver normally sees the drab imma-

ture fish resting on the bottom where they somewhat resemble a blenny.
The body is rather long and the head is flattened with the eyes set almost on top of the head. There is a strong branched spine at the angle of the front gill cover and this is important in identifying the species. The pelvic fin is inserted in front of the pectoral fin.
These are bottom living fishes of shallow water. The gill opening is situated high up on the body and plumes of mud may be blown out of it; presumably as part of the feeding process. The adult males often have strikingly elongated dorsal fin rays and brilliant colouring.

Callionymus belenus Risso   p. 240

Mediterranean.

Body slender; *head* and front part of body flattened from above; *eye* large (diameter greater than the space between the eyes); *mouth* small; *jaws*, upper jaw slightly longer than lower; spines, 3 spines present on a small bone situated on the front gill cover; *fins*, 2 dorsal fins, the 1st small with 3 rays; 2nd fin longer. In the male the last ray of the 2nd dorsal fin is branched and taller than the rest. Tail fin rounded with no elongated rays. 1D III; 2D 8-9; A 8.
Greyish-yellow with large pale spots and small dark spots. 1st dorsal fin is entirely black in the female and immature males and with a black and white spot in the adult males.
Up to 8 cm.

Found on sand and mud from 15-150 m. and probably much deeper. Breeding takes place in the spring and summer.

Callionymus phaeton Valenciennes   p. 240

Mediterranean and adjacent Atlantic.

Body slender; *head* flattened from above; *snout* pointed; *eye* very large; *mouth* reaches back to the front margin of the eye; *lips* fleshy; *spines*, 2 spines on the bone on the front gill cover; *fins*, 2 dorsal fins. All the rays of the 2nd dorsal and tail fins are branched. In the adult male the last ray of the 2nd dorsal fin and the centre ray of the tail fin are elongated as is also the last ray of the anal fin which is also branched. The female and immature fish have no elongated rays and the 2nd dorsal fin is shorter. 1D IV;

2D 8-9; A 8-9.
The colour of this fish is very variable, usually orange or pinkish with greenish spots on the back and sides and pinkish silver on the under surface. The dorsal fins are reddish. The 1st dorsal fin has a black spot between the 3rd and 4th rays. There are faint darker spots on the dorsal fins and the anal fin has a darker lower edge. The lower half of the tail fin has dark spots. Pectoral and pelvic fins are reddish.
Females up to 12 cm. Males up to 18 cm.

A fairly deep water fish found between 100-500 m. on mud or sand bottoms. They feed on small crustacea and worms etc. which they find on the bottom.

Dragonet   p. 240 **pl. 89, 90**
Callionymus lyra L.

Mediterranean; East Atlantic from Norway and Iceland to Senegal; North Sea; English Channel; west Baltic.

Body slender; *head* and front half of body flattened from above. The head appears triangular when seen from above; *snout* long; *jaws*, upper jaw longer than the lower; *lips* fleshy; *spines* there are 4 spines present on a bone on the front gill cover; *skin* without scales; *fins*, in the adult male the 2 dorsal fins are large and the 1st spine of the 1st fin is greatly elongated and reaches beyond the base of the tail fin. The anal fin is also large. In the female and immature male the 2nd dorsal and anal fins are smaller than in the adult male. The 1st dorsal is very much smaller and is lower than the 2nd dorsal. The pectoral and pelvic fins are joined by a small membrane.
1D IV; 2D 9; A 9.

Callionymus belenus   p. 239

3 front gill cover spines;
1st dorsal fin short with 3 rays.

Adult male last ray of dorsal fin branched and elongate;
1st dorsal fin with a black and white spot.

Female and immature male: 1st dorsal fin entirely black.
8 cm.

Callionymus phaeton   p. 239

2 spines on front gill cover;
body pinkish with green spots;
black spot on 1st dorsal fin.

Adult male: elongate last rays of 2nd dorsal fin, anal fin and centre ray of tail fin.
18 cm.

Females 12 cm.

Dragonet *Callionymus lyra*   p. 239

4 front gill cover spines.

Adult male: very elongate 1st ray of 1st dorsal fin;
yellow and blue dorsal fins.
30 cm.

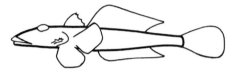

Female and immature male: uniform fin colour;
6 greenish or brownish blotches along sides.
20 cm.

Callionymus reticulatus   p. 241

3 front gill cover spines;
bluish spots on body;
10 rays in 2nd dorsal fin.

Adult male: dark spots and wavy blue lines on dorsal fins.
10 cm.
Females 8 cm.

Spotted Dragonet *Callionymus maculatus*

p. 242

Dorsal fins with alternating dark and pearl spots;
4 front gill cover spines.

Males 14 cm.

Females 11 cm.

Callionymus festivus p. 242
3 front gill cover spines.

Adult male: 1st dorsal fin shorter than 2nd; wavy bluish lines along the fins and blue spots and stripes along body.
15 cm.

Female and immature male: 1st dorsal fin black;
colour similar to adult male but paler.
10 cm.

Colour extremely variable and may depend on the individual, the locality and the season. There is a considerable difference between the male and female. Male fish are usually yellowish or brownish with bluish striping and spotting on the head and body. The 1st dorsal fin is yellow with blue blotches and the 2nd dorsal is yellow with longitudinal blue stripes. The females and young males are predominantly yellowish-brown with 6 greenish or brownish blotches along the sides and 3 dark saddles across the back. The fins are uniform in colour.
Females up to 20 cm. Males up to 30 cm.

These fish are very common in some areas and may be found from very shallow water to 50 m. and sometimes below 100 m. They usually live on sand or mud into which they sometimes bury themselves. They feed on small molluscs, worms and crustacea which they find buried in the sand.

Spawning occurs from January through to the end of August, but the month varies from one locality to another. The fish move into more shallow water prior to spawning where the male displays his large fins to the female. The female, if she accepts the male, swims to his side and spawning takes place in mid water with the fish swimming side by side nearly vertically upwards towards the surface.

Male fish live for about 5 years and females for about 7 years.

Callionymus reticulatus Valenciennes p. 240

English Channel; Irish Sea; western Mediterranean.

This species is very similar to *Callionymus maculatus* except that there are 3 spines on the front gill cover bone and always 10 rays in the 2nd dorsal fin. Mature males have the 1st spine of the 1st dorsal fin elongated and the last ray of the 2nd dorsal fin branched.

Adult females and immature males are orange-brown on the back and sides and off-white below. There are bluish spots and

4 clearly defined orange-red saddles on the back. Adult males are similar in colour but the dorsal fins have rows of dark spots and bluish wavy lines.
Males up to 10 cm. Females up to 8 cm.

These fish are easily mistaken for *Callionymus maculatus* and may well be more common than records suggest. They are normally found from 20-40 m. on sand or shell bottoms.

Spotted Dragonet   p. 241
Callionymus maculatus Rafinesque

East Atlantic from Norway to Gascony; south Iceland; English Channel; Mediterranean.

Body slender; *head* somewhat flattened from above; *spines* there are 4 spines present on a small bone situated on the front gill cover; *skin* without scales; *fins* in the male the 2 dorsal fins are enlarged; the 1st ray of the 1st dorsal fin reaches to the end of the base of the 2nd dorsal. The 1st dorsal is separated from the 2nd. In the female the fins are shorter and the bases of the 2nd dorsal and anal fins are larger. 1D IV; 2D 9-10; A 8-9.
The colour of both the males and the females are similar. The ground colour is brownish-yellow darker on the back and lighter on the belly. There are longitudinal rows of dark and pearly spots. Male fish have four dark spots interspaced with pearl spots on each membrane between the rays of the 2nd dorsal fin. The 1st dorsal and tail fin are also spotted. The female fish have 2 rows of dark spots on the 2nd dorsal fin. The spots are less distinct in the female and immature male than in the adult male.
Females up to 11 cm. Male up to 14 cm.

Found from shallow water to below 300 m., usually on sandy bottoms. They are frequently found on the large offshore sandbanks. They are carnivorous and feed on small bottom living animals, worms, crustacea etc. In the Atlantic they spawn from April to June and in the Mediterranean from January to May.

Callionymus festivus   p. 241

Mediterranean.

Body slender; *snout* rather long and rounded; *mouth* small; *lips* fleshy; *skin* without scales; *spines*, 3 spines present on a bone situated on the front gill cover; *fins*, in the male the 2nd dorsal fin is higher than the 1st and all the rays are elongated beyond the margin of the membrane. The last ray of the anal fin and all the rays of the tail fin are similarly elongated and the two central rays of the tail fin are longer than the others. The female fish has none of the elongated rays and the 2nd dorsal and anal fins are more equal in size. 1D IV; 2D 7; A 9.

The ground colour is greyish, sandy or brownish. In the adult male there are bluish spots and stripes on the back and sides of the body and on the head. The fins are yellow with wavy bluish lines outlined in dark brown. The lower edge of the tail and anal fins are blackish. The female is similar to the male but less bright, also the 1st dorsal fin is black.
Male up to 15 cm. Female up to 10 cm.

These fish live in very shallow water on sand. They may bury themselves in the sand and can even travel through it to emerge a little distance further off. Breeding occurs during summer. The male displays to the female with his large fins and circles her in gradually decreasing circles. The male and female swim close together and spawning occurs in an almost vertical position when swimming towards the surface.

Sub-order: AMMODYTOIDEA

Perciform fishes with elongated bodies and no spiny rays in the fins.

Family:

AMMODYTIDAE

The **Sand Eels** can be extremely abundant especially in more northern waters over sandy bottoms. They are extremely important as they provide one of the main sources of food for the carnivorous fishes that are eaten by man. Their great abundance is not always evident for they spend much of the time buried in the sand and their silvery camouflage is so perfect that divers have reported seeing only a host of disembodied eyes moving through the water where, in reality, a whole school was present. The Sand Eels have elongated eel-like bodies with a single soft-rayed dorsal fin set in a groove along the back. The upper jaw which is extensible is shorter than the lower. The tail fin is forked and free from the dorsal and anal fins.

The schools are tight and well-disciplined and give a shimmering appearance underwater as each fish is swimming with an eel-like motion.

Sand Eels Ammodytidae   **pl. 192**
     pl. 54

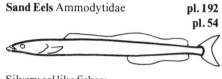

Silvery eel like fishes;
swim in shoals.
About 20 cm.

Smooth Sand Eel

Gymnammodytes semisquamatus (Jourdain)

East Atlantic south to Biscay; English Channel; North Sea.

Body long and slender; *snout* pointed; *jaws*, lower jaw longer than upper. Upper jaw protrusible; *lateral line* has short cross canals running at right angles from the main canal. For each canal running upwards there are 2 directed downwards; *scales* small only present on the rear of the body; *grooves*, there is a groove running below the pectoral fin which is only slightly longer than the fin; *fins* 1 long dorsal fin which starts behind the pectoral fin and has 2 shal-

low dips along its length. Anal fin less than 1/2 the length of the dorsal fin and has 1 shallow dip. Tail fin forked. D 56-59; A 28-32; vertebrae 65-70.

Back golden or greenish brown. Yellowish sides and silver underneath. These fish appear silver and are extremely well camouflaged with only the eye being conspicuous. Up to 23.5 cm.

Usually found from 20-200 m. often on shell or gravel bottoms in which they often bury themselves. They are often encountered swimming in large shoals. Spawning occurs in the winter in the English Channel and progressively later northwards so that in the North Sea spawning occurs in late spring and summer. The eggs are laid on the sea bed. The adults feed on very small crustacea and worms.

Sand Lance

Gymnammodytes cicerellus (Rafinesque)

Mediterranean.

This species was for many years confused with the Atlantic species the Smooth Sand Eel *Gymnammodytes semisquamatus*. The

Sand Lance is, however, the only common Sand Eel in the Mediterranean and may be distinguished by the lower cross canals of the lateral line which are very short. D 53-59; A 27-32.

Back greenish, silvery flanks and belly. There is a blue-black spot on the head.

Up to 16 cm.

Found in large shoals over, or in, sand. Spawns in winter.

Sand Eel
Ammodytes tobianus L.
= *Ammodytes lancea* Cuvier

East Atlantic south to Gibraltar; English Channel; North Sea; Iceland; west Baltic; occasionally Mediterranean.

Body long and slender; *snout* pointed; *jaw*, upper jaw protrusible, lower jaw longer than upper; *lateral line* without any cross canals; *scales* over whole body and are situated in diagonal skin folds and appear in regular rows. There are patches of scales on the base of each tail fin lobe; *fins* dorsal fin starts in front of the tip of the pectoral fin. There are no dips in the dorsal and anal fins; *grooves* the groove below the pectoral fin reaches past the beginning of the anal fin. D 50-56; A 25-31; vertebrae 60-66.

Yellowish green or bluish on back, yellowish sides, silver belly.

Up to 20 cm.

Very common from shallow water to 30 m. Usually found over sand in large shoals. Spawning occurs during the winter in the English Channel and the eggs are laid amongst the sand. In the Baltic there are thought to be 2 races, one spring spawning, the other autumn spawning. They feed on small fish and worms.

Sand Eel
Ammodytes marinus Raitt

East Atlantic south to English Channel north beyond Iceland; North Sea; west Baltic.

This species is very similar to *Ammodytes tobianus* but may be distinguished by the scales on the belly which are not in regular rows; no scales at the base of the tail fin lobes and a greater number of rays in the dorsal and anal fins. D 55-67; A 26-35; vertebrae 66-72.

Back greenish-blue, flanks bluish silver, belly silver.

Up to 25 cm.

These fish are not found as shallow as *Ammodytes tobianus*, usually not above 30 m. but are very common. Spawning occurs during winter in the southern part of their distribution and early spring in their northern range off Iceland. They feed on small worms and very small crustacea, larvae and eggs. Commercially important as they are a food source for herring and are also eaten by sea birds.

Greater Sand Eel
Hyperoplus lanceolatus (Lesauvage)

East Atlantic; English Channel; North Sea; west Baltic.

This fish may be most easily identified by the black spot on the sides of the snout. Other characters are the scales which are not situated in skin folds; 2 teeth on the roof of the mouth and the upper jaw which can swing forward but is not truely protrusible. D 52-61; A 28-33; vertebrae 65-69.

Back bluish-green or brownish-green, sides bluish green, belly silver. There is a dark spot on the sides of the snout.

Up to 32 cm., it is the largest of the Sand Eels.

These fish are found from shallow water to 150 m. usually over or in sand. Breeding occurs in water between 20-100 m. deep during spring and summer. They feed on small crustacea, larvae, eggs and small fishes.

Hyperoplus immaculatus Corbin

West coast of the British Isles; German North Sea coast.

D 59-61; A 31-34; vertebrae 70-74.
Up to 30 cm.

This fish is easily confused with the Greater Sand Eel *Hyperoplus lanceolatus* and it may be more extensive than records suggest.
It may be identified by the uniformly dark snout and the greater number of vertebrae.

Back bluish-green, sides lighter, belly silvery. The snout is dark.
Very little is known about this species. They are caught on sand, gravel and shell bottoms both near and offshore. Spawning occurs in winter and spring.

Family:

BLENIIDAE

The **Blennies** generally live in very shallow water even between tide marks. They superficially resemble the gobies but do not have the pectoral fins united into a sucker. When alarmed they often take refuge in holes or crevices amongst rocks and this may explain why some Blennies (e.g. *Blennius rouxi*) are quite often seen by divers but are rare in museum collections.

They are elongated fishes without scales and with a somewhat slimy body. There is a single dorsal fin which is usually divided into the hard and soft rayed portions by a notch. The pelvic fins are reduced and contain only 1-3 rays. The hard rays are flexible and jointed at the base. Many species have little 'tentacles' on the head and these are particularly useful in identifying blennies from photographs.

Butterfly Blenny　　　　　　**pl. 98**
Blennius ocellaris L.　　　　　　p. 246

Mediterranean and north-east Atlantic from Senegal to northern Scotland.

Body deep and laterally flattened; *teeth* there are 2 prominent canine teeth at the back of each jaw preceeded by smaller rounded teeth; *tentacles* there are 2 rather squat branched tentacles above each eye and two very small ones on each side of the 1st dorsal ray; *fins* the dorsal rays are very long, the 1st being the longest. There is a notch separating the two parts of the dorsal fin.
D X-XII, 14-16; A I-II, 15-16.
Greenish brown with 5-7 darker vertical bands on the flanks extending onto the dorsal fin. There is a characteristic white spot surrounded by a pale-bluish halo on the 6th and 7th rays of the dorsal fin.
Up to 20 cm.

Range extends into deeper water than most other blennies. In northern Europe it is generally found on *Lithothamnium* or shell bottoms from 10-20 m. In the Mediterranean it may be as deep as 100 m. and is often found in discarded pieces of crockery etc.
Spawns in late winter in the Mediterranean, in spring and early summer in northern Europe. The eggs are laid on the sheltered inside face of mollusc shells or other suitable objects and are guarded by the male. The larvae are planktonic but the young fish settle on the bottom after a few weeks.

Red-speckled Blenny　　　　p. 246 **pl. 95**
Blennius sanguinolentus Pallas

Mediterranean and north east Atlantic from Senegal to the English Channel.

Body heavy with a paunchy stomach. The front part of the body is laterally flattened;

245

Butterfly Blenny *Blennius ocellaris*   p. 245

High dorsal fin with a large light edged black spot.
20 cm.

Red-speckled Blenny   p. 245
Blennius sanguinolentus

Body heavy and paunchy;
massive conical teeth.
15 cm.

Tompot Blenny *Blennius gattorugine* p. 248

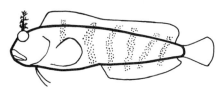

Large branched appendage above eye.
20 cm.

Horned Blenny *Blennius tentacularis*   p. 248

Large appendage above eye fringed on rear edge only.
15 cm.

Zoanimir's Blenny *Blennius zvonimiri* p. 248

Stagshorn-like appendage above eye;
smaller appendages on each nostril.
6 cm.

Blennius cristatus   p. 249

Tuft of 7 filaments above each eye;
line of filaments along mid line of back
from between eyes to the dorsal fin;
profile rounded.
10 cm.

Peacock Blenny *Blennius pavo*   p. 249

Black spot with blue margin behind eye;
dark bands on flanks outlined in blue;
10 cm.

Zebra Blenny *Blennius basilicus*   p. 249

Maze-like pattern on cheeks.
12 cm.

Blennius sphinx   p. 249

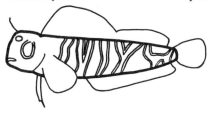

Red-bordered spot on gill cover;
simple appendage above eye.
8 cm.

Striped Blenny *Blennius rouxi*   p. 250

Black stripe on flanks.
7 cm.

Shanny *Blennius pholis*   p. 250

No appendages on head;
Atlantic only.
16 cm.

Blennius trigloides   p. 250

No appendages on head;
Mediterranean only.
9 cm.

Blennius canevae   p. 250

Golden cheeks.
7 cm.

Black-headed Blenny *Blennius nigriceps*
                                     p. 251

2 dorsal fins; red body;
black patterned head;
no appendages on head.
4 cm.

Montagu's Blenny *Blennius galerita*   p. 251

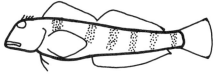

Fringe of branched appendages between
eyes;
row of simple appendages on mid line of
head between eyes.
8 cm.

Blennius dalmatinus   p. 251

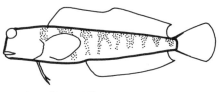

Outline of dorsal fins;
no tentacles above eyes.
4 cm.

profile rounded; *jaws* rather small; *teeth* there are two rather small canine teeth in each jaw behind files of massive conical teeth; *tentacles* there is a minute tentacle above each eye about 1/3 diameter of the eye; *fins* the dorsal fin is of about the same height throughout its length.   D XII-XIII; A 22-24.

Colour variable, usually dull olive. There are often red spots on the tail and pelvic fin rays and often a black spot on the 1st and 2nd dorsal rays. There is often a dark margin to the anal fin.

Up to 15 cm., sometimes 20 cm.

Lives in shallow water amongst rocks where there are plenty of crevices. They are voracious carnivores using their strong teeth to crush up molluscs and echinoderms. They breed in early summer laying their eggs in empty mollusc shells etc.

Tompot Blenny   p. 246
Blennius gattorugine Brünnich   **pl. 91, 92**

Mediterranean and north east Atlantic from West Africa to northern Scotland. Not in North Sea.

Body rather elongated and laterally flattened behind; *eyes* set high on the head; *profile* rounded; *mouth* large extending back to the centre of the eye; *teeth* there is a single row of minute sharp teeth in each jaw. No canines; *tentacles* there is a large much-branched tentacle above each eye. There is also a minute fringed tentacle on the nostril beneath each eye; *fins* the dorsal fin is continuous but the last rays are the longest. D XII-XIV, 17-20; A I, 20-21.

Ground colour yellow brown, olive brown or reddish brown. There are about 7 darker vertical bars on the flanks extending onto the dorsal fin. Eye is reddish brown.

Up to 20 cm., sometimes 30 cm.

In the northern part of its range it is found in the kelp (*Laminaria*) zone but in S. England and in the Mediterranean it may live 1 m. or less beneath the surface. They favour places where there are medium-sized stones. They are very common in the Eastern Channel.

Spawns in early spring. The male guards the eggs which are laid in crevices between rocks. The male also keeps them oxygenated by fanning a current of water over them. After a month the eggs hatch. The larvae are free-swimming but probably settle on the bottom by mid-summer.

Horned Blenny   p. 246
Blennius tentacularis Brünnich   **pl. 93**

Mediterranean and North East Atlantic from Senegal to Portugal.

Body similar to Tompot Blenny (*Blennius gattorugine*) but the body is somewhat longer and less massive. The most obvious point of difference is in the *tentacles* above each eye which are branched on the rear edge only. On each of the front nostrils there is a small tubular tentacle with a tiny tongue-like projection on the rear part of its tip. *fins* D XII-XIV, 18-23; A II, 21-24.

Colour variable. The most common form is chestnut brown and pale ochre arranged in alternating blotches; the body is speckled with darker brown. The iris is reddish with blue spots. There is often a black spot on the 2nd dorsal ray.

Up to 15 cm.

Generally lives on sandy bottoms where there are some stones at depths from 6-30 m. They are also found in brackish water on mud.

Zoanimir's Blenny   p. 246
Blennius zvonimiri Kolombatovic

Mediterranean.

Body rather flattened laterally; *eye* set high in the head; *profile* almost vertical in front of the eye; *mouth* small not extending past the hind edge of the eye. Lips fleshy, particularly the upper; *teeth* 2 large canines in each jaw behind a single file of small conical

teeth; *tentacles*, above each eye there is a prominant tentacle resembling a stag's antlers and on each nostril there are smaller horned tentacles; *fins*, dorsal is obviously divided into two halves. D XII, 17; A I, 19.
Grey or red-brown with darker marbling. There are red spots and stripes on the sides of the head and on the lips.
Up to 6 cm.

Uncommon or rare. They live in water of moderate depth where there is a rich growth of encrusting red algae.

Blennius cristatus L.   p. 246

Both sides of tropical Atlantic, rarely in Mediterranean.

Profile steep and rounded; *tentacles*, there is a fringe of filaments running along the mid line of the back from between the eyes to the dorsal fins and a tuft of 7 filaments above each eye; *fins* there is a small notch in the dorsal fin at the 12th ray. D XI-XII, 13-16; A I-II, 15-18.
Olive green with a dark spot between the 1st and 2nd dorsal rays.
Up to 10 cm.

Rare. In shallow water, probably lives amongst rocks.

Peacock Blenny   p. 246
Blennius pavo Risso

Mediterranean.

Body rather long and laterally compressed; *eye* very small; *profile* steep and rounded. In the mature male there is a large hump above the eye; *mouth* very small set low in the head; *teeth* a single file of blunt teeth in each jaw. At the rear of each file there is a single canine tooth; *tentacles* very small, one above each eye; *fins* the dorsal fin is continuous, not obviously divided into two parts. D XII, 21-23; A 24-46.
The ground colour is yellow-green with darker vertical bands. The front ones are

outlined in iridescent blue. At the rear the blue pattern breaks down into scattered spots. There is always a black spot edged with iridescent blue behind the eye. In the female the black spot may be bordered with pinkish white. In the mature male the hump has patches of golden brown.

Up to 10 cm.

In very shallow water, usually on mud and sand near rocks. Very tolerant of extremes of temperature and salinity and may enter water that is nearly fresh.
Breeds in summer. They deposit their eggs amongst stones or on empty shells.

Zebra Blenny   p. 246
Blennius basilicus Valenciennes

Mediterranean.

Resembles the female of the Peacock Blenny but differs in having violet vertical bars on the flanks and a maze-like pattern of dark markings outlined in white on the sides of the head.
Up to 12 cm.

Very rare, shallow coastal water.

Blennius sphinx Valenciennes   p. 247
                                                        pl. 97
Mediterranean.

Body long and laterally compressed; *profile* very steep; *eye* small and set high in the head; *mouth* very small set low in the head; *teeth* two canine-like teeth in each jaw behind a single file of small sharp teeth; *tentacles* there is a conspicuous unbranched tentacle above each eye; *fins*, the dorsal fin is clearly divided into two parts. D XII, 16-17; A 18-20.
Ground colour yellowish-brown with irregular dark vertical bands outlined in whitish blue; behind the eye there is a blue-grey spot which is outlined in red.
Up to 8 cm.

Generally lives in very shallow water where the bottom is rocky but the seaweed is not thick. Rests with its high dorsal fin folded over to one side, but when alarmed the fin becomes erect.

Spawns in early summer in rock cavities, shells etc. The courtship is typical of Blennies. The male displays in front of any passing female following and butting her until she enters the nest. A single male may mate with a succession of females. The eggs are attached to the nest wall with sticky filaments.

Striped Blenny p. 247 **pl. 94**
Blennius rouxi Cocco

Mediterranean.

Body long; *eyes* large; *profile* almost vertical; *mouth* very small; *tentacles*, above each eye there is a tentacle made up of 5 or 6 filaments. There is also a very small tentacle on the lower nostrils; *fins*, dorsal fin not obviously divided. D XI, 24; A II, 24.
Pearly white with a black longitudinal stripe running from the eye to the tail. The fins are colourless except that there is sometimes a black spot between the 1st and 2nd dorsal ray and the anal fin has a black margin.
Up to 7 cm.

Normally lives in depths between 1 and 7 m. where there are rocks, some detritus and possibly some sand. They are often seen with their heads looking out of tiny holes in the rocks. Considered to be quite a rare fish, although divers often see it, perhaps because of its very conspicuous colouration.

Shanny p. 247 **pl. 96**
Blennius pholis L.

North East Atlantic south to Portugal.

Body distance of snout to hind edge of eye equals distance of hind edge of eye to the 1st dorsal ray; *profile* steep but not vertical; *mouth* rather large; *teeth* two canine teeth in each jaw behind a single row of small

sharp teeth; *tentacles* no tentacle over eye; *fins* a shallow notch separates the two parts of the dorsal fin. D XI-XIII, 18-20; A I, 17-19.
Colour is blotched greenish or yellowish brown with an indistinct spot between the first and second parts of the dorsal fin. During courtship and nesting the male becomes intense black with white lips.
Up to 16 cm.
Common amongst rocks and stones in shallow water, often in the intertidal. Also found on sandy or muddy shores where they will hide beneath fronds of seaweed.
The Shanny feeds on almost anything including barnacles and algae. Spawning is in early summer. The eggs are usually laid on the underside of rocky crevices and are guarded by the male which also fans them to keep them oxygenated. A long-lived fish, a large blenny of 16 cm. may be 16 years old.

Blennius trigloides Valenciennes p. 247

Mediterranean.

Resembles the Shanny except that the distance between the tip of the snout and the hind edge of the eye is greater than the distance of the hind edge of the eye to the first dorsal ray. D XII, 16; A 17-18.
Up to 9 cm.

Blennius canevae Vinciguerra p. 247

Mediterranean.

Head rounded; *eye* very small set high in the head; *mouth* set low in the head; *teeth*, two canines in each jaw followed by a single row of conical teeth; *tentacles*, none on head; *fins* the two parts of the dorsal fin are divided by a deep notch. D XIII, 16; A II, 15-16.
Cheeks and throat are golden brown speckled with red spots. The body colour is dark brown covered with an irregular network of lighter lines arranged in a more or less longitudinal fashion.
Up to 7 cm.

Uncommon, lives amongst rocks in water so shallow that the fish is exposed by the waves.

Black-headed Blenny p. 247
Blennius nigriceps Vinciguerra

Mediterranean.

Body elongate; *eyes* large set high in the head; *profile* rather steep; *mouth* set very low; *appendages* none on head; *teeth* 2 well developed canines in each jaw; *fins* no predorsal fin (see *Tripterygion* spp.). The two parts of the dorsal fin are conspicuously divided by a deep notch. D XII, 15; A I, 15.
Up to 4 cm., perhaps the smallest fish in our area.

Greatly resembles *Tripterygion minor* in general colour but differs in having no predorsal fin. Lives in the same habitat as *Tripterygion minor*.

Montagu's Blenny p. 247 **pl. 99**
Blennius galerita L.
= *Blennius montagui* Fleming

North East Atlantic from Guinea to south Ireland; rarely in Mediterranean.

Body long; *profile* gently sloping; *mouth* large; *teeth* 2 well developed canines in the lower jaw; *tentacles*, there is a characteristic crest of fringed filaments running between the eyes. From the centre line of this is a row of 3-7 simple filaments running back along the mid-line of the head; *fins*, there is a conspicuous notch between the two parts of the dorsal fin. D XII-XVIII, 5-18; A 17-18.
Olive brown often with darker vertical bands. In young fish there are bluish-white spots on the head and body. In the males the crest is orange and longer than in the females. The body colour is also darker.
Up to 8 cm.

Chiefly lives in rock pools and rarely extends down to the *Laminaria* seaweed zone. They live in pools where there is little algal growth and mostly feed on the arms of barnacles.
Breeds in mid-summer in rock pools. The male guards the eggs which are attached to the ceiling of a rock crevice with adhesive filaments.

Blennius dalmatinus Steindachner & Kolombatovic p. 247

Mediterranean.

Body long and laterally compressed; *profile* almost vertical; *tentacles* there are no tentacles above the eye but there is a pair of exceedingly small simple tentacles above the lower nostrils; *jaws* set very low; *teeth*, 20-22 small almost conical teeth in each jaw. On each side of the lower jaw at the back is a larger tooth; *fins* the spiny rays of the dorsal fin are shorter than the soft rays. D XII, 15-16; A II, 18-19.
Olive-green above, greenish-gold below. There are 8-11 darker brown bands on the back which may be outlined in silver. In some specimens the dark bands are divided along a line running from the gill cover to the tail. On the top of the head is a dark area which is divided from the greenish-gold cheeks by a wavy line.
Up to 4 cm.

Before the advent of divers this little fish was thought to be very rare as it can only be caught in a hand net. However, it may, in reality, be quite common. It lives on rocky bottoms where there is little weed from just beneath the surface to a maximum of 2 m.

Family:

CLINIDAE

Only one member of this family is present in our area. Elongated fishes with tiny scales embedded in the skin. The pelvic fins contains only one or two rays.

Cristiceps argentatus (Risso)

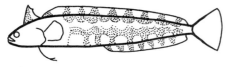

2 dorsal fins the 1st between eye and gill cover.
10 cm.

Mediterranean.

Body long with a sharp snout; *mouth* small and terminal with fleshy lips; *fins*, there are 2 dorsal fins. 1st dorsal is small and triangular set very far forward at the level of the front gill cover. 2nd dorsal very long and of uniform height. The outer rays of the tail are fused in pairs, the inner rays are single

and widely spaced. 1D III, 2D XXV-XXVIII, 3-4; A II, 18-20.
The ground colour can be almost any shade of black, olive, violet or brown. The flanks have rows of lighter blotches and the dorsal and anal fins have regular dark vertical bands.
Up to 10 cm.

Lives near rocks especially where there is Sea Grass (*Posidonia*). They rarely swim but instead crawl over the bottom on their pectoral fins. Feed on the small animals coating the fronds of Sea Grass and algae.
The eggs are laid in early summer in about 1 m. of water. The eggs are provided with tufts of silky filaments and these are used to attach the eggs to algal fronds. The larvae do not have a free living stage but immediately settle on the bottom.

Family:

TRIPTERYGIIDAE

Blenny-like fishes with the body covered with scales. Three dorsal fins of which the first can be erected independantly of the others.

Black-Faced Blenny   p. 253
Tripterygion tripteronotus

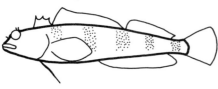

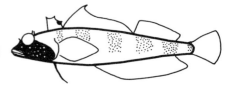

3 dorsal fins;
tentacles over eye;
indistinct dark bars on body;

Adult male black head;
elongate 1st and 2nd rays of 2nd dorsal fin.
8 cm.

Black-Faced Blenny p. 252 **pl. 101, 102, 103**
Tripterygion tripteronotus (Risso)

Mediterranean.

Body rather long; *profile* snout pointed; the forehead slopes at an angle of 67° and the profile between the eye and snout is concave; *mouth* is terminal, the lower lip does not appear pendulous; *scales* cover the body except for the gill cover; *tentacles*, there is a simple tentacle above each eye and another on each of the front nostrils; *fins*, there are three dorsal fins, 1st dorsal small set far forward at the level of the front gill cover. The 2nd dorsal has 1-3 rays elongated into filaments in the adult males. 1D III; 2D XVII-XVIII; 3D 12-13; A 24-27.

Sexes differ in colour. The females are a drab mottled brown. The males are red or orange with a black head. Both sexes have indistinct dark vertical bands on the body. There are sky-blue spots on the cheeks and there are blue edges to the 2nd and 3rd dorsal fins. During spawning the head of the male becomes a more intense black.
Up to 8 cm.

Found on rocky shores from a few centimetres to 10 m. An agile species that does not have the usual blenny habit of seeking protection in holes and crevices but it does frequent places protected from direct sunlight. Breeds in early summer. The males defend a territory about 1 m. across. The female spawns with a succession of males.

Pigmy Black-Faced Blenny   **pl. 104**
Tripterygion minor (Risso)

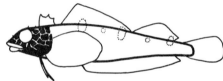

black head;
red body;
usually 5 white spots along back.
5 cm.

Mediterranean.

Resembles the Black-Faced Blenny (*Tripterygion tripteronotus*) but is smaller, has a sharper snout (the forehead slopes at an angle of 58°) and the lower lip has a pendulous appearance.
No marked difference between the sexes. The head is black and the body is red with three large and two small white spots on the back.
Up to 5 cm.

Tend to frequent rock areas near the mouth of dark crannies and crevices especially where there are encrusting red algae. It is sometimes accompanied by *Blennius nigriceps* to which it shows a very strong superficial resemblance. The reason for the association does not seem to be known.

Family:

PHOLIDIDAE

Body compressed with very small scales; vertical fins run together, dorsal fin long and low with 75-100 spines. No lateral line. Pelvic with 1 spine and 1 small soft ray.

Butterfish; Gunnel   **pl. 120**
Pholis gunnellus (L.)
= *Centronotus gunnellus* Schneider

Row of 9-13 black spots circled with white along the base of the dorsal fin.
25 cm.

Eastern Atlantic south to the English Channel, north beyond Iceland; English Channel; North Sea; west Baltic.

Body long and flattened from side to side; *head* small; *snout* rounded; *lips* fleshy; *scales* very small; *lateral line* absent; *fins* 1 very long dorsal fin which consists of small spines and commences just behind the gill cover and continues to the base of the tail. 1 anal fin which is about half the length of the dorsal fin. Pelvic fins very reduced and consist of 1 spine and 1 small soft ray. D LXXV-LXXXII; A II, 39-45.
Greyish-brown or brown; there are irregular darker vertical bands on the body but these gradually break up to form a mottled pattern in older fish. There is a row of 9-13 black spots surrounded by a white ring situated along the back at the base of or on the

dorsal fin. There is also a dark stripe which runs from the eye to the rear edge of the mouth.
Up to 25 cm.

These fish are common and found from the sea shore down to 40 m. and live in a variety of habitats from mud and sand to rocks, frequently amongst kelp (*Laminaria*) holdfasts. They feed on small molluscs and crustacea, worms and other small sedentary animals. Breeding takes place in winter and nests may be made from between tides to 25 m. The egg masses are guarded by either parent though more usually by the female. During this period both the parent fish cease feeding. The skin is very slimy (hence the name Butterfish) and they are extremely difficult to catch.

Family:

LUMPENIDAE

Very long rather rounded body covered with small scales, lateral line indistinct or absent, tail fin free with 13 principle rays. Pelvics with 1 spine and 3-4 branched rays.

Yarrell's Blenny pl. 100
Chirolophis ascanii (Walbaum)

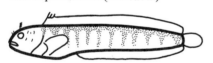

Branched tentacles above each eye;
1 long dorsal fin;
pelvic fin small.
25 cm.

Western and southern coasts of Iceland, Norway and Britain, central east coast of England, western English Channel.

Body rather deep behind the head and tapers to the tail fin; *snout* rounded; *tentacles*, there is a large branched tentacle above each eye and a smaller tentacle at the rear edge of each nostril. There are small fila-

ments present on the top of the head; *scales* very small, not present on the head; *fins*, dorsal fin long. It originates behind the head and extends to the tail. The first few spines may be extended into short filaments. These are particularly apparent in adult males. Pelvic fins small (shorter than dorsal fin height). D L-LXV; A I, 35-40.
Yellowish or greenish-brown with darker vertical bands. The eye is circled with dark brown and there is also a dark stripe which runs from the eye to the corner of the mouth.
Up to 25 cm.

Usually found from 40-50 m., occasionally down to 400 m., amongst rocks. Very little is known about this fish except that it feeds on molluscs, worms etc.

Family:

STICHAEIDAE

Long-bodied blenny-like fishes, body usually scaled; lateral line present, usually 15 principle rays in the tail fin. Pelvics when present have normally branched soft rays.

Snake Blenny; Lumpenus
Lumpenus lumpretaeformis (Walbaum)

Eel-like body;
row of dark spots along flanks;
tail fin long and slender;
long pelvic fin;
no tentacles.
42 cm.

Northern Atlantic south to Denmark

Body long and slender, rather eel-like; *snout*
rounded; *scales* very small; *fins*, dorsal fin

long with the first and last few rays shorter
than the rest. Pelvic fins soft, slender and
longer than the dorsal fin rays. Tail fin long
and slender. D LXVIII-LXXVI; A I, 47-53.
Back and flanks pale brown, pale green or
pale blue with dark brown irregular
blotches. Under surface greenish-yellow except under the head and front part of the
body which is bluish.
Up to 42 cm.

Usually found from 40-80 m., also down to
150 m., on mud, sand or stony areas.
Spawning occurs in the winter and young
fish take up a bottom-living existence at
about 3 months. Adults feed on very small
crustacea, and echinoderms.

Family:

ZOARCIDAE

The **Eel Pouts** are elongated blenny-like fish with a long dorsal and a shorter anal fin forming a continuous fringe around the tail. The pelvic fins are set well in front of the pectorals.
The head is massive.

Eel Pout; Viviparous Blenny   pl. 105
Zoarces viviparus (L.)

Dorsal fin continues around tail;
distinct dip near tail.
46 cm.

Northern Irish Sea; North Sea; eastern
Channel; Baltic; Scandanavian coasts to the
White Sea.

Body long, tapering towards the tail; *teeth*
conical; *scales* very small embedded in the
skin which is slimy; *fins*, 1 dorsal fin which
is continuous around the tail with the anal
fin. There is a distinct dip in the dorsal fin
near the tail. Pelvic fins small. D 72-85 +
V-XI + 16-24; A 80-95.
Yellowish green or brown on the back shading to greyish-brown underneath. There are
darker diffuse spots and bands on the back
and head and a row of blotches along the
sides. There is a row of dark arches on the
dorsal fin. The pectoral fins are edged with
yellow or orange; but during the breeding
season they may be bright red in the male.

255

Up to 46 cm.

Found in very shallow water from 4-10 m., occasionally down to 40 m. They usually live on sand or, mud into which they may bury themselves or amongst seaweed and under stones. These fish are viviparous, fertilisation of the eggs is internal. The young fish, up to 300 at any one time, develop in the single ovary and are born after about 4 months. They feed on crustacea, molluscs, small fish and other small animals.

Family:

ANARHICHADIDAE

The **Wolf Fish** or **Cat Fish** are blenny-like fishes with no pelvic fins, a long dorsal fin and a distinct tail fin. There are large canine-like teeth in the front of the jaws and flattened grinding teeth at the sides.

Spotted Catfish; Lesser Catfish	**Wolf Fish; Cat Fish** pl. 195
Anarhichas minor Olafsen	*Anarhichas lupus* L.

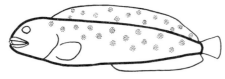 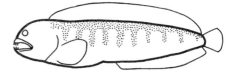

Large conical teeth;
spots scattered over back and flanks.
200 cm.

Northern East Atlantic south to Scotland; Iceland; Greenland; Barents Sea; White Sea; also western Atlantic south to Maine.

Body deepest behind the head tapering strongly to the tail; *head* large; *snout* rounded; *teeth* large, curved teeth in the front of the jaws and rounded or pointed teeth on the palate and the sides of the jaws; *fins* 1 dorsal fin and 1 anal fin which terminates nearer the tail than the dorsal fin. No pelvic fins. D 74-80; A 45-47.
Brownish with distinct darker spots irregularly scattered over the back and sides.
Up to 200 cm.
Found from 25-460 m. but most commonly between 100-250 m. They live mainly on mud or sand. During the spring there is an offshore migration into deeper water for spawning which occurs at around 250 m. The eggs are laid in large clumps. The adult fish feed on small molluscs, echinoderms, crustacea and fish.

Large curved teeth in front of jaws, flattened teeth at sides;
dark vertical stripes on the back and flanks.
120 cm.

Eastern Atlantic south to the English Channel, North Sea, Iceland, Scandanavia, Greenland also western Atlantic south to New Jersey.

Body, deepest behind the head, tapering to the tail; *head* large; *teeth* curved in the front of the jaws; large, flattened teeth at the sides; *fins* 1 dorsal fin and 1 anal fin which terminates nearer the tail than does the dorsal fin. No pelvic fins. D 69-79; A 42-48.
Greyish, brownish-red or greenish with darker vertical stripes which continue onto the dorsal fin.
Up to 120 cm.
Usually found from 100-300 m. but young fish may be seen in much shallower water. Spawning occurs in winter in water between 40-200 m. The eggs are laid in clumps on the sea bed amongst weeds and stones. The adult fish feed on small echinoderms, molluscs, crustacea etc.

Family:

OPHIDIIDAE

Generally deep water eel-like fishes with pelvic fins resembling barbels set far forward.

Snake Blenny   pl. 106, 193   *Parophidion vassali* (Risso)
Ophidion barbatum L.   = *Ophidion vassali* Risso

Pelvic fins resembling barbels under chin; united median fins with dark margin.
30 cm.

Mediterranean and East Atlantic from Biscay to Senegal.

Body long, and eel-like, laterally flattened; *head* one fifth to one sixth the total body length; *scales* are an elongate oval shape and are embedded in the skin on the upper surface of the fish in a mozaic pattern; *eye*, large having a diameter about 1/4 the head length; *fins*, the dorsal, anal and tail fins are united into a single continuous fin. The pelvic fins are set very far forward under the chin; each consists of a double ray and superficially resemble barbels. D 125-140; A 115-120.
Pinkish above, paler bluish-white below. The united fin has a continuous dark margin.
Up to 30 cm.

Found on sandy or muddy bottoms down to about 150 m. The young may be found in the vicinity of sea grass (*Posidonia*) meadows.

Pelvic fins resembling barbels under chin; united median fins without dark margins.

Superficially resembles *Ophidion barbatum* except that the united fin has no dark rim. The most important anatomical difference is in the swim bladder which is simple in structure in *P. vassali* but very complex in *Ophidion barbatum*.
D 127-135; A 100-110.
A rare, little-known fish. Probably lives much deeper than *Ophidion barbatum*.

Family:

CARAPIDAE

The **Pearlfish** have elongated bodies with long dorsal and anal fins merging at the tail. No tail or pelvic fins. The anus opens far forward near the pectoral fin. They live in the body cavity of sedentary echinoderms and molluscs.

Pearlfish

Pearlfish
Carapus acus (Brünnich)
= *Fierasfer acus* Kaup

Long, slender shape;
no tail or pelvic fins;
vent opens in front of pectoral fin;
lives in sea cucumbers.
20 cm.

Mediterranean.

Body long, slender and laterally flattened grading gradually in depth from the head to the pointed tail; *jaws* large, extending back past the eye; *vent*, situated in front of the pectoral fin; *scales* none; *fins*, dorsal and anal fins run most of the length of the body and unite at the tail. In the early larval stages the 1st ray of the dorsal fin is much elongated. No pelvic or tail fins.

D approx. 140; A approx. 170
Silver white with reddish spots and mottlings.
Up to 20 cm.

The Pearlfish inhabits the body cavity of sea cucumbers especially *Stichopus regalis* and *Holothuria tubulosa*. The young enter head first through the anus but the adults enter tail first. Inside they feed mostly upon the gonads of their host which occasionally eviscerates expelling the fish. At intervals the body of the Pearlfish pulsates in its host presumably to create a current of oxygenated water inside. The Pearlfish is able to live freely outside the body of a sea cucumber but soon finds a new host which it recognises by its mucus and by its long shape.
Breeding occurs in summer, the eggs form gelatinous aggregations; the young larvae hatch after about three days and develop in the plankton.

Echiodon drummondi Thompson
= *Fierasfer dentatus* Cuvier (Kaup)

Resembles *Carapus acus* except that the vent is situated slightly behind rather than slightly in front of the pectoral fin.
North-east Atlantic from Norway to Biscay. A related species *Echiodon dentatus* (Cuvier) lives in the Mediterranean usually below 100 m.

Gobies
by Peter J. Miller, Department of Zoology, Bristol University

Sub-order: GOBIOIDEA

Family:

GOBIIDAE

Gobies belong to the suborder Gobioidea (Gobioidei), a large and very varied array of chiefly tropical and warm-temperate fishes, which in habitat are predominantly inshore marine but which also include many estuarine and freshwater species. In addition to over forty kinds of goby found in the Mediterranean and adjacent eastern Atlantic, the Eurasian fauna contains a further thirty or more brackish and freshwater species restricted to the Black and Caspian Seas and their drainage systems. Only the former group can be dealt with here.

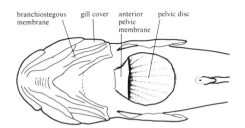
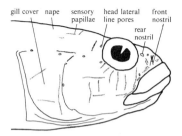

Rock Goby *Gobius paganellus* showing some characters in
Goby classification

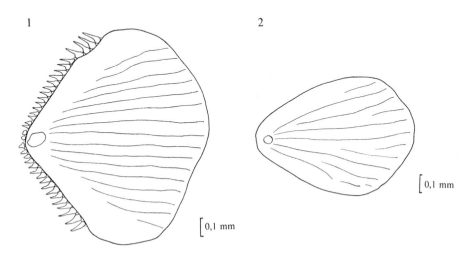

Two types of scale: 1. Ctenoid (from tail stalk of Zebra
Goby, *Zebrus zebrus*); 2. Cycloid (from tail stalk of Banded
Goby, *Chromogobius quadrivittatus*)

259

These European members of the family Gobiidae are typically small fishes, all less than 30 cm. in length, of moderately elongate and nearly cylindrical or compressed body shape, with a rounded, somewhat depressed head, displaying prominent cheeks and dorso-lateral eyes, the latter being separated by a relatively narrow space. The mouth-cleft is oblique, and the snout short, with a tubular front nostril, sometimes bearing a triangular lappet or from one to several narrow tentacles on its free rim, and a generally pore-like rear nostril (page 259). There are two dorsal fins, the first of merely six spinous rays in most species, a single anal fin, and a tail fin whose rear outline is usually rounded. The pectoral fins are large and, in some species (*Gobius*), the uppermost pectoral rays are more or less free from the fin-membrane, so that their branches form a fringe along the upper edge of the fin (see Recognition Chart to Larger Gobies). An important external characteristic of gobiid fishes is the fusion of the pelvic fins to provide a shallow funnel-like disc whose front (anterior) rim is completed by a transverse membrane between the spinous rays of each pelvic fin. This pelvic disc has weak suctorial powers. The anterior membrane may display lateral lobes (in some *Gobius* species; p. 268) or carry minute papillae (villi) along its free edge (in some *Pomatoschitus* spp.; p. 275). As a secondary modification in certain gobies (e.g. *Gobius auratus, Odondebuenia balearica*), the pelvic fins become more or less separated again, with at least notching or cleaving of the posterior rim of the disc and reduction or loss of the pelvic anterior membrane (pp. 269, 272).

Unlike most fishes, the gobies do not possess a lateral line along each side of the body but canals belonging to this sensory system occur on the head and there are rows of exposed neuromast sensory papillae over the head and body (p. 259). The precise arrangement of these lines of papillae and the degree of reduction of the head canals are of considerable value in goby classification, but such features are often difficult to examine without chemical treatment of the dead fish or use of a binocular microscope. Their description has therefore been omitted from the following account.

European gobies of marine occurrence fall clearly into two groups according to habitat. Most species spend their juvenile and adult life on or near the sea-bed, with merely the postlarval stage dwelling in the plankton. Three genera (*Aphia, Pseudaphya, Crystallogobius*) remain pelagic over the entire life-span (apart from breeding activities), and retain gobiid larval features such as body elongation and transparency. In diet, all the European species are predatory, feeding chiefly on tiny crustaceans, worms, molluscs, and sometimes small fish. Their dentition consists typically of a few rows of fine teeth in each jaw, and similar pharyngeal teeth. In certain species, the lower jaw carries lateral canine teeth and some enlarged teeth may be found posteriorly situated in the middle of the upper jaw.

Gobies breed during spring and summer months. Adult males delimit territories about a nest and spawning occurs after courtship display and conduction of a ripe female into the nest. Nests are excavated under stones, shells or other rigid objects (tiles, old shoes, etc.), and the eggs deposited in a patch on the underside of this roofing structure. Each adult female may be expected to produce several batches of eggs in the course of one breeding season. Goby eggs are pear-shaped or more elliptical, not exceeding 3-4 mm. in length among the present species, and are attached to the substrate by tiny filaments radiating from the base of each egg. The eggs are guarded and fanned by the male until hatching. The postlarvae (released with yolk sac already absorbed) live in the plankton for a few weeks or months before adopting a bottom-living existence (unless permanently pelagic in mode of life). Longevity in gobies ranges from one to several years, according to species. The pelagic forms, *Aphia minuta* and *Crystallogobius linearis*, have been cited as 'annual' vertebrates, adults dying at the age of about one year immediately after their first breeding season. Smaller species, such as those of the genus *Pomatoschistus*, become sexually mature after their first winter of life. Larger, longer-lived *Gobius* species, at least in northern waters, do not mature until two years old. A maximum age of ten years has been recorded in a female Rock Goby (*G. paganellus*) by counting the number of transparent annual rings laid down in her earstones (otoliths).

The following systematic treatment of European marine gobies omits a few rare species or those whose specific nomenclature and diagnosis require further investigation. In some cases, new English common names have had to be coined for Mediterranean species. The number of scales in lateral series is counted along the lateral midline from pectoral axilla ('armpit') to base of tail fin. In dealing with fin-rays, the last ray in both the second dorsal and anal fin, always bifurcate, is counted as one. Maximum length cited always refers to total length and is thus inclusive of the tail fin. Most of the gobies mentioned in the text are also shown in diagrammatic lateral outline to indicate certain major features useful for identification. Other details of coloration, etc., are omitted. Except in a few cases, specimens illustrated are adult males.

There is still much to be learned about the systematics, general biology, and distribution of gobies, and the amateur naturalist can help to solve these problems by field observations and collection of specimens. The present author is always glad to receive gobies for identification from any part of Europe.

Larger Gobies
(*Gobius, Zosterisessor, Thorogobius*)

Black Goby *Gobius niger*   p. 266   **Slender Goby** *Gobius geniporus*   p. 269

Black spots in front corners of dorsal fins; elongate first dorsal rays; lateral scales 32-42.
15 cm.

Body slender; lateral scales 50-55.
16 cm.

Bucchich's Goby *Gobius bucchichii*   p. 267   **Red-mouthed Goby** *Gobius cruentatus*

p. 269

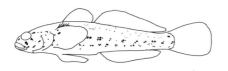

Many small spots on head and body; lateral scales 50-56.
10 cm.

Lips and cheeks with red streaks; black lines of sensory papillae on head; lateral scales 52-58.
18 cm.

Rock Goby *Gobius paganellus*   p. 267

Pale to orange band on upper edge of first dorsal fin; free pectoral rays well developed; lateral scales 50-57.
12 cm.

Grass Goby *Zosterisessor ophiocephalus*
p. 272

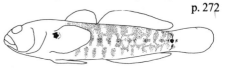

No free pectoral rays; greenish, with irregular dark bars; small pectoral and tail spots; lateral scales 53-65.
25 cm.

Giant Goby *Gobius cobitis*   p. 268

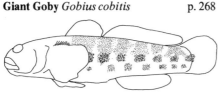

Pelvic membrane with lateral lobes; lateral scales 59-67.
27 cm.

Leopard-spotted Goby   p. 271
Thorogobius ephippiatus

Large brick-red blotches; no free pectoral rays; lateral scales 33-42.
13 cm.

Medium-sized Gobies
(*Lesueurigobius, Buenia, Deltentosteus*)

Sanzo's Goby *Lesueurigobius sanzoi*   p. 273

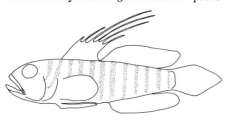

Vertical dark and yellow bands; scales on nape; lateral scales 25-26; first dorsal rays very elongate in male.
9.5 cm.

Jeffrey's Goby *Buenia jeffreysii*   p. 273

Four black spots on lateral midline; nape naked; lateral scales 25-30.
6 cm.

Fries' Goby *Lesueurigobius friesii*   p. 273

Yellow spots; scales on nape; lateral scales 28-29.
10 cm.

Le Sueur's Goby *Lesueurigobius suerii*
p. 273

Blue and yellow markings on head, body and fins; nape naked; lateral scales 26-28.
5 cm.

262

Four-spotted Goby p. 274
Deltentosteus quadrimaculatus

Black spots along lateral midline; first dorsal fin with dark markings and spots; scales on nape; lateral scales 33-35.
8 cm.

Deltentosteus colonianus p. 274
As for *Deltentosteus quadrimaculatus* but angle of mouth below rear half of eye.
8 cm.

Medium to small Gobies
(*Pomatoschistus, Gobiusculus*)

Note that pectoral fin has been omitted in all diagrams except that of female *P. microps*.

Sand Goby *Pomatoschistus minutus* p. 275

Anterior pelvic membrane with villi; lateral scales 58-70; males with about four vertical bars, one distal first dorsal spot, and pale breast.
9.5 cm.

Pomatoschistus norvegicus p. 276

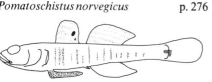

Pelvic villi; lateral scales 55-58; males with more numerous vertical striae.
6.5 cm.

Quagga Goby *Pomatoschistus quagga* p. 277

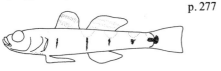

Scales in lateral series 35-40; vertical bars in both sexes, but not more than four present.
5 cm.

Painted Goby *Pomatoschistus pictus* p. 276

Dorsal fins with rows of large black spots and rosy bands; lateral scales 36-43; dusky breast.
5.7 cm.

Kner's Goby *Pomatoschistus knerii* p. 277

Tail fin square-cut; lateral scales 38-46; vertical bars in both sexes (more numerous in males), and large tail spot.
4 cm.

Marbled Goby *Pomatoschistus marmoratus* p. 276

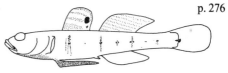

Anterior pelvic membrane with villi; lateral scales 40-46; males with about four vertical bars, more or less distal first dorsal spot, and dusky breast.
6.5 cm.

Common Goby *Pomatoschistus microps*
p. 276

Scales in lateral series 39-52; males with numerous vertical bars, single first dorsal spot (nearer base of fin), and dusky breast and throat.
6.4 cm.

Female without lateral bars and merely inconspicuous first dorsal spot; breast pale.

Tortonese's Goby *Pomatoschistus tortonesei*
p. 277

Scales in lateral series 29-35; lateral bars in both sexes.
3 cm.

Two-spotted Goby *Gobiusculus flavescens*
p. 275

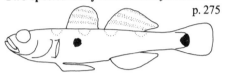

First dorsal fin with seven rays; large caudal spot; males with conspicuous lateral spot below first dorsal fin. Lateral scales 35-40.
6 cm.

Smaller Gobies
(*Gobius, Corcyrogobius, Odondebuenia, Zebrus, Chromogobius, Didogobius, Lebetus*)

Striped Goby *Gobius vittatus*   p. 269

Broad dark stripe along head and body; lateral scales about 36.
4 cm.

Liechtenstein's Goby   p. 272
Corcyrogobius liechtensteini

Dark spot on each side of throat; lateral scales 27-31.
2.5 cm.

Diminutive Goby *Lebetus scorpioides*
p. 274

No anterior pelvic membrane; lateral scales 25-29; male with large dusky yellow first dorsal fin.
4 cm.

Female with purplish brown lateral markings; dark spot on first dorsal fin.
4 cm.

Coralline Goby *Odondebuenia balearica*
p. 272

Pelvic fins separate; modified scales at tail fin base; lateral scales 24-32.
3.2 cm.

Zebra Goby *Zebrus zebrus*   p. 270

Free pectoral rays; lateral bars; scales in lateral series 33-37.
4.5 cm.

Kolombatovic's Goby   p. 271
Chromogobius zebratus

Five broad pale saddles; edge of pectoral bar acutely bent; lateral scales 41-52.
5.5 cm.

Ben-Tuvia's Goby *Didogobius bentuvii*
p. 271

Pale, somewhat translucent; eyes relatively small; tail fin elongate; lateral scales 65-70.
5 cm.

Banded Goby   p. 270
Chromogobius quadrivittatus

10-14 verical dark bars; two pale saddles; intense spot in lower angle of gill cover; edge of pectoral bar straight; lateral scales 56-72.
6.5 cm.

Gobies not illustrated include:

Large-scaled Goby *Thorogobius macrolepis*
p. 271
Pale spots on head, no free pectoral rays; lateral scales 27-28.
6 cm.

Roule's Goby *Gobius roulei*   p. 266
No scales on nape. Elongate first dorsal rays. Lateral scales 33-34.
8 cm.

Sarato's Goby *Gobius fallax*   p. 268
Pelvic discs deeply cleft. Brownish olive colour. 3 pairs intense black spots on abdomen. Lateral scales 39-45.
9 cm.

Golden Goby *Gobius auratus*   p. 268
Canary-yellow colour. Pelvic disc deeply cleft. Lateral scales 44-45.
10 cm.

Bellotti's Goby *Gobius ater*   p. 267
Dark brown. Lateral scales about 40.
7 cm.

Schmidt's Goby *Gobius Schmidti*   p. 270
Pelvic disc cleft; lateral scales 50-53. Life colour unknown.
6.5 cm.

Pelagic Gobies
(Aphia, Pseudapha, Crystallogobius)

Transparent Goby *Aphia minuta*   p. 278

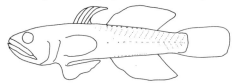

Five first dorsal fin-rays; lateral scales 24-25; adult male with larger teeth and fins.
5.1 cm.

Female with minute teeth; small pelvic disc.
4.6 cm.

Ferrer's Goby *Pseudaphya ferreri*   p. 278

Large tail spot; five first dorsal fin-rays; lateral scales 25-26.
2.9 cm.

Crystal Goby *Crystallogobius linearis*
                     p. 278

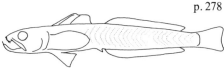

First dorsal fin with not more than two rays; no scales; males with enlarged canine teeth and deep pelvic disc.
4.7 cm.

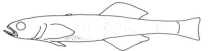

Females without teeth, small pelvic disc, and often no first dorsal fin.
3.9 cm.

Black Goby   p. 261 **pl. 109, 110**
Gobius niger L.
= *Gobius jozo* L.

Mediterranean and Black Sea; Eastern Atlantic, from Cape Blanco, Mauretania, to Trondheim, Norway, including Baltic Sea.

Fins Pelvic anterior membrane without lateral lobes; pelvic disc rounded (not emarginate); pectoral fins with short free upper rays; 1st dorsal fin elongate in the male, reaching to at least the middle of the 2nd dorsal fin when depressed. *Nostrils* Front nostril with a simple flap on rim. *Scales* The

nape is scaled but to a variable extent. *Fin rays* 1D, V-VII (usually VI): 2D I, 11-13 (usually 12-13); A I, 10-13 (usually 11-12). Scales in lateral series 32-42 (usually 35-41).

Coloration pale brownish, with darker lateral blotches and dots, to dusky in breeding males; each dorsal fin with dark spot in upper anterior corner; branchiostegous membrane dark.
Length 15 cm.

Found in coastal waters down to 50-75 m., on sandy or muddy ground, often in seagrass meadows and, being tolerant of brackish water, also penetrating estuaries and lagoons. Breeds from March-May (Mediterranean), May-August (Baltic). Often sold in Mediterranean fish-markets.

Roule's Goby   p. 265
Gobius roulei De Buen

North-western Mediterranean.

Fins Pelvic anterior membrane and disc complete, with no lateral lobes; the pectoral fin has free rays; rays of the first dorsal fin elongate, with the 4th ray reaching to about the middle of the second dorsal fin-base when depressed. *Scales* No scales on nape; 33-34 scales in lateral series. *Nostrils* Front nostril with a simple triangular lappet on rim. *Fin rays* 1D, VI; 2D, I, 12; A, I, 11.

Coloration in life unknown.
Length 8 cm.

Inshore, but biology otherwise unknown. The validity of this species is open to doubt and it may prove to be synonymous with the Black Goby, *G. niger.*

Bucchich's Goby   p. 261 **pl. 116**
Gobius bucchichii Steindachner

Mediterranean and Black Sea.

Fins Pelvic anterior membrane and disc complete but without lateral lobes. The uppermost rays of the pectoral fin are free. *Nostrils* Front nostril with a simpel tentacle, sometimes forked. *Scales* Nape scaled;

scales in lateral series 50-56. *Fin rays* 1D VI; 2D I, 13-14 (usually 14); A I, 12-14 (usually 13).

Coloration fawn or darker brown, with longitudinal rows of numerous small dark spots along head and body, usually more noticeable on lateral midline and below.
Length 10 cm.

Occurs on inshore sandy and muddy grounds, in the vicinity of the Snakelocks sea-anemone *Anemonia sulcata*, among whose tentacles the goby seeks refuge when alarmed.

Bellotti's Goby   p. 265
Gobius ater Bellotti
= *Gobius balearicus* Lozano y Rey

North-western Mediterranean.

Fins Pelvic anterior membrane without lateral lobes but otherwise well developed; the pelvic disc is rounded (not emarginate). Pectoral fins with free upper rays. *Nostrils* The front nostril has a divided tentacle on the rim. *Scales* No scale on nape; about 40 scales in the lateral line series. *Fin rays* 1D VI; 2D I, 12-14; A I, 11.

Coloration dark brown, with dusky fins.
Length 7.1 cm.

This is a little-known species, probably of inshore occurrence. Its biology is unknown, and full geographical distribution remains to be traced.

Rock Goby   p. 262 **pl. 111**
Gobius paganellus L.

Mediterranean and Black Sea; Eastern Atlantic, from West Africa to southern and western shores of British Isles.

Fins Pelvic anterior membrane well developed with lateral lobes absent or showing only feebly. Free rays of the pectoral fin very conspicuous, reaching to the origin of the 1st dorsal fin (at least in small individuals). *Nostrils* Front nostril with tentacles divided into as many as six or even more finger-like branches. *Scales* The nape is

scaled (and there are scales on the upper rear region of the cheek in *Gobius punctipinnis* Canestrini, which is possibly a variety of *G. paganellus*). Scales 50-57 in lateral series. *Fin rays* 1D VI; 2D I, 13-14; A I, 11-12.

Coloration fawn, with darker mottling and lateral blotches, to dark; upper margin of first dorsal fin with pale horizontal band. Ripe males deep purplish brown, with dorsal band yellow or orange.
Length 12 cm.

Inshore shallows on rocky ground in Mediterranean. Along Atlantic coasts, the rock goby is common under stones in pools on sheltered rocky shores with abundant weed cover. Breeding season January-June (Naples), April-June (Isle of Man).

Giant Goby   p. 262 **pl. 119**
Gobius cobitis Pallas
=*Gobius capito* Cuvier & Valenciennes

Mediterranean and Black Sea; Eastern Atlantic from Morocco to western mouth of English Channel.

Fins Pelvic anterior membrane with conspicuous lateral lobes, pelvic disc short and rounded; free upper pectoral rays; front nostril with a tentacle which is usually divided into several finger-like processes of

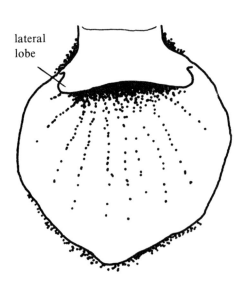

lateral lobe

variable length. *Scales* Nape scaled; 59-67 scales in lateral series. *Fin rays* 1D VI; 2D

I, 13-14 (usually 13); A I, 10-12 (usually 11).
Coloration brownish-olive 'pepper and salt' speckling with dark mottling and rounded blotches along and below lateral line, the latter especially distinct in small individuals. Ripe males dark, their median fins with narrow white edges.
Length 27 cm.—the largest Mediterranean-Atlantic goby.

Found on rocky and weedy ground in shallows. On the south-west coast of England, this species has been found only in higher shore pools, where the water may be brackish. Breeds from March to May (Mediterranean). Sold in Mediterranean fish-markets.

Sarato's Goby   p. 265
Gobius fallax Sarato

Mediterranean.

Fins Pelvic anterior membrane little developed and disc deeply cleft; pectoral fins with free pectoral rays. *Nostrils* Front nostril with a simple triangular flap. *Scales* Nape and upper corner of gill cover scaled; 39-45 scales in the lateral line series. *Fin rays* 1D VI, 2D I, 12-14 (usually 14); A I, 11-13 (usually 12-13).

Coloration brownish-olive, with mottling and lateral blotches; usually three pairs of intense but small dark spots visible on abdomen in ventral view.
Length almost 9 cm.

Little is known concerning the biology of this species, which is probably of inshore occurrence.

Golden Goby   p. 265
Gobius auratus Risso

Western Mediterranean; records from English Channel apply to a related species, which still requires description.

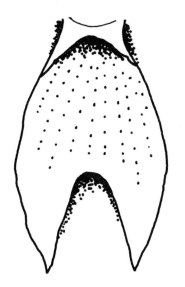

Fins Pelvic anterior membrane little developed and pelvic disc deeply cleft. Upper pectoral rays free. *Nostrils* Front nostril with a slight triangular flap. *Scales* Nape and upper part of gill cover scaled; 44-45 scales in lateral line series. *Fin rays* 1D VI; 2D I, 14; A I, 13.

Coloration canary-yellow; small dark spot at upper end of pectoral fin base.
Length 10 cm.

Inshore, on rocky ground with algae and gorgonians, from 15-80 m. Found a short distance above sea-bed. Breeding season not recorded.

Striped Goby   p. 264
Gobius vittatus Vinciguerra

Mediterranean.

Fins Pelvic disc with slight cleft (emarginate) and anterior membrane almost lacking. *Nostrils* Front nostril with a simple process on rim. *Scales* Nape scaled; about 36 scales in lateral series. *Fin rays* 1D VI; 2D I, 11-12; A I, 11.

Coloration golden yellow with broad dark brown to black stripe along upper part of flank from snout, through eye, to tail fin.
Length 4 cm.

Offshore, 15-85 m., but may live above sea-bed. Biology otherwise unknown, although the conspicuous lateral stripe suggests that this species may participate in 'cleaning symbiosis' with other fishes. Similar marking is shown by tropical Western Atlantic Neon Gobies, *Elacatinus*, which are known to practice this mode of feeding.

Slender Goby   p. 261   **pl. 113**
Gobius geniporus Valenciennes

Western Mediterranean.

Body slender, maximum depth being about seven times total length. *Fins* Pelvic anterior membrane well developed although lacking lateral lobes; pelvic disc with straight (truncate) rear edge; pectoral fins with merely the tips of the upper rays free from the membrane. *Nostrils* Front nostril with a simple or lobe-like process on rim. *Scales* Nape scaled; 50-55 scales in the lateral line series. *Fin rays* 1D VI; 2D I, 12-13; A I, 11.

Coloration brownish with dark lateral blotches; fins dusky.
Length 16 cm.

Inshore, on sand or mud, around sea-grass meadows. Breeds from April to May (Taranto).

Red-mouthed Goby   p. 261   **pl. 112**
Gobius cruentatus Gmelin

Mediterranean; Eastern Atlantic, from North Africa to southern Ireland.

Fins Pelvic anterior membrane well developed but without lateral lobes; pelvic disc with slightly cleft (slightly emarginate) rear edge; free pectoral rays. *Nostrils* Front nostril with a simple tentacle. *Scales* Present on nape, upper part of gill cover, and on the rear part of the cheek. 52-58 scales in later-

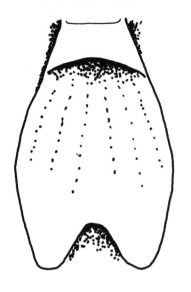

al series. *Fin rays* 1D VI; 2D I, 14; A I, 12-13.

Coloration reddish brown with lateral blotches; lips and cheeks display vivid red markings; rows of sensory papillae on head are black.
Length 18 cm.

Inshore, 15-40 m., on rocky and sandy ground or in sea-grass meadows.

Schmidt's Goby   p. 265
Gobius schmidti (De Buen)
= *Gobius assoi* De Buen

Western Mediterranean, including Adriatic Sea; recently reported from Black Sea.

Fins Pelvic anterior membrane reduced and delicate, connecting merely the bases of the pelvic spinous rays; pelvic disc with cleft rear margin (emarginate); upper pectoral rays free only at tips if at all. *Nostrils* Front nostril with a small triangular flap. *Scales* Nape scaled; 50-53 scales in lateral series. *Fin rays* 1D VI; 2D I, 14; A I, 13.

Coloration in life unknown.
Length 6.5 cm.

Found on coralline grounds, 25-40 m. Biology otherwise unrecorded.

Zebra Goby   p. 265
Zebrus zebrus (Risso)

Mediterranean.

Fins Pelvic disc complete, but anterior membrane without lobes; pectoral fin with well-developed free rays. *Nostrils* Front nostril with a tentacle as long as the nostril tube; rear nostril a low tube near the eye. *Scales* None on nape; 33-37 scales in lateral series. *Fin rays* 1D VI; 2D I, 11; A I, 8-10.

Coloration brownish olive, with several vertical dark bands across flanks; pale band across anterior nape behind eyes, continuing very obliquely backwards to upper gill cover; pectoral base with dark band, having deeply concave front edge; first dorsal fin with two dark bands, upper edge reddish.
Length 4-4.5 cm.

Inshore shallows, inshore pools, or in weedbeds.

Banded Goby   p. 265
Chromogobius quadrivittatus (Steindachner)
= *Relictogobius kryzhanovskii* Ptchelina

Mediterranean and Black Sea.

Fins Pelvic anterior membrane complete, with small lateral lobes; pelvic disc short, with a straight rear edge and rounded corners (rounded truncate); pectoral fins without free rays; all scales cycloid (see p. 259). *Nostrils* Front nostril without process on rim; rear nostril a short tube. *Teeth* Canines in sides of lower jaw; enlarged rear canines in middle of upper jaw. *Scales* No scales on nape; 56-72 scales in lateral series. *Fin rays* 1D VI; 2D I, 8-11; A I, 7-9.

Coloration pale brown with 10-14 vertical dark bars across flanks; pale saddle at origin and end of second dorsal fin; broad pale band across nape and root of pectoral fin, with its posterior edge sharply demarcated by straight black band across pectoral base; head, cheeks and gill cover with convolute pattern; intense black spot in lower front corner of gill cover.
Length 6.5 cm.

Inshore and intertidal.

Kolombatovic's Goby   p. 265
Chromogobius zebratus (Kolombatovic)

Adriatic Sea, and eastern Mediterranean.

Similar body-form to the Banded Goby, *Ch. quadrivittatus*, but original scales ctenoid and only 41-52 in lateral series.

Coloration with five broad pale saddles across back; cheek with simpler pattern not extending onto branchiostegous membrane and no black gill cover spot on; edge of pectoral band bent.
Length nearly 5.5 cm.

Biology unknown; probably on offshore coralline grounds.

Ben-Tuvia's Goby   p. 265
Didogobius bentuvii Miller

Eastern Mediterranean, off coast of Israel.

Body Elongate, laterally compressed. *Eyes* Small and widely separated. *Fins* Tail-fin very long and lanceolate; pelvic anterior membrane and disc complete; pectoral fins without free rays. *Teeth* Lower jaw with a large lateral canine tooth on each side; upper jaw with large median rear teeth. *Nostrils* Front nostril somewhat tubular, overlying upper lip; no rim process. *Scales* All cycloid, 65-70 in lateral series *Fin rays* 1D VI; 2D I, 14; A I, 12.

Coloration somewhat pale and translucent in life, pale fawn preserved with numerous melanophores over head and body.

Length almost 5 cm. (only one specimen so far described).

The single type-specimen was trawled on muddy sand in about 37 m.

Leopard-spotted Goby   p. 262   **pl. 114**
Thorogobius ephippiatus (Lowe)
= *Gobius forsteri* Corbin

Mediterranean; Eastern Atlantic, from Madeira to south-western and western shores of the British Isles.

Fins Pelvic disc and anterior membrane complete, without lateral lobes; no free rays in pectoral fin; tail fin rounded. *Nostrils* Front nostril has no process. *Scales* None on nape; 33-42 in lateral series. *Fin rays* 1D V-VII (usually VI); 2D I, 10-12 (usually 11); A I, 10.

Coloration pale fawn with brick-red blotches (dark on preservation) over head and body, including lateral mid-line; black spot in rear corner of first dorsal fin.
Length 13 cm.

Coastal, in or near crevices associated with vertical rock faces, from Low Water Spring Tide to 30-40 m.; occasionally in shore pools. In captivity, this species is recorded as breeding from May-July (Plymouth).

Large-scaled Goby   p. 265
Thorogobius macrolepis (Kolombatovic)

Recorded only from Adriatic Sea (Split, Jugoslavia).

As preceding species but pelvic disc slightly emarginate and anterior pelvic membrane lacking or hardly developed. *Scales* in lateral series 27-28; *fin-rays* 1 D VI; 2D I/ 11; AI, 10.

Coloration in life not recorded in detail; small pale spots on head instead of large brick-red blotches as found in Leopard-spotted goby.
Length about 6 cm.

Biology unknown.

Grass Goby   p. 262 **pl. 115**
Zosterisessor ophiocephalus (Pallas)
= *Gobius lota* Valenciennes

Mediterranean and Black Sea.

Fins Pelvic disc and anterior membrane complete but with no lateral lobes; pectoral fin without free rays. *Nostrils* No tentacle on rim. *Scales* Present on nape; 53-65 in lateral series. *Fin rays* 1D VI; 2D I, 14-15; A I, 13-15.

Coloration yellowish olive to greenish, with many irregular vertical brownish bars; blotches along lateral midline, with one at base of tail more intense, and sometimes small dark spot at upper end of pectoral fin-base.
Length 25 cm.

Brackish water, in estuaries and lagoons, on mud and eel-grass beds; especially common in Venetian lagoon. Breeds from April-May (Black Sea). Sold in fish-markets.

Liechtenstein's Goby   p. 264
Corcyrogobius liechtensteini (Kolombatovic)

Known only from the Adriatic island of Korcula (Jugoslavia).

Fins Pelvic anterior membrane and disc complete but with lateral margins slightly elongated at the rear (emarginate posteriorly); no villi or lateral lobes on anterior membrane; no free pectoral rays; first dorsal fin elongate in males, reaching to rear half of

the second dorsal fin-base. *Nostrils* Front nostril without process on rim. *Scales* 27-31 (usually 27-29) in lateral series. *Fin rays* 1D VI; 2D I, 9; A I, 7-8 (usually 8).
Coloration in life not described; an intense black spot on the underside of the head on the branchiostegous membrane of each side.
Length 2.5 cm.

Biology unknown; possibly an inhabitant of coralline grounds.

Coralline Goby   p. 265
Odondebuenia balearica (Pellegrin & Fage)

Mediterranean; Eastern Atlantic, on Morroccan coast.

Fins Pelvic fins virtually separate and therefore the anterior membrane is absent; no

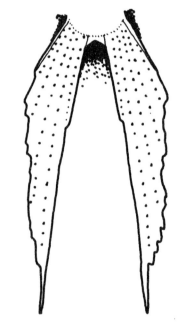

free rays in the pectoral fin; first dorsal fin elongate, reaching to at least the middle of the second dorsal fin in males. *Scales* None on nape. Uppermost and lowermost scale on each side of the tail fin is enlarged, with very long lateral ctenii (see p. 259); 24-32 (usually 28-30) scales in lateral series. *Fin rays* 1D VI; 2D I, 9-10 (usually 10); A I, 9-10 (usually 9).

Coloration of male blood-red, with narrow vertical blue bands across flanks; females without vertical bars.
Length 3.2 cm.

Found on coralline grounds 25-70 m.

Fries' Goby p. 262
Lesueurigobius friesii (Malm)

Mediterranean; Eastern Atlantic, from Spain to eastern North Sea.

Fins Pelvic anterior membrane without villi or lobes; pelvic disc complete; no free pectoral rays; 1st dorsal rays elongate; tail fin lanceolate. *Scales* Cover nape; 28-29 scales in lateral series. *Fin rays* 1D VI; 2D I, 13-16; A I, 12-15.

Coloration pale fawn with numerous golden yellow dots over nape, body, dorsal and tail fins.
Length 10 cm.

Offshore, on mud or muddy sand, usually over 50 m.; probably in association with Norway Lobster, *Nephrops norvegicus*, in whose burrows the gobies may conceal themselves. Breeding season March-July (British Isles).

Le Sueur's Goby p. 262
Lesueurigobius suerii (Risso)
= *L. lesuerii* (Risso)

Mediterranean and Eastern Atlantic (Morocco)

Fins Pelvic anterior membrane without villi or lobes and disc complete; no free pectoral rays; 1st dorsal ray somewhat elongate (especially in males) reaching back to the rear part of the second dorsal fin. Tail fin lan-

ceolate with bluntly pointed tip.
Nostrils Front nostril simple. *Scales* Nape without scales; 26-28 scales in lateral series. *Fin rays* 1D VI; 2D I, 13; A I, 13.

Coloration of body with vertical markings of blue and yellow; head and gill cover exhibit oblique yellow bands on blue background. Yellow lines on dorsal and tail fins.
Length over 5 cm.

Inshore species; young offshore. Breeding season summer and autumn (Mediterranean).

Sanzo's Goby p. 262
Lesueurigobius sanzoi (De Buen)

Mediterranean; Eastern Atlantic, Madeira to southern Spain.

Fins Pelvic anterior membrane without villi or lobes; pelvic disc complete; no free pectoral rays; tail fin lanceolate; 1st dorsal fin rays very elongate in male, extending when depressed to the end of the second dorsal fin or root of tail fin. *Nostril* Front nostril simple. *Scales* Present on nape; 25-26 scales in lateral series. *Fin rays* 1D VI; 2D I, 15; A I, 16-17.

Coloration of vertical dark brown and yellow bands across flanks.
Length over 9.5 cm.

Offshore, muddy sand or mud, 50 m. Biology otherwise unknown.

Jeffrey's Goby p. 262
Buenia jeffreysii (Günther)

Eastern Atlantic, Celtic Sea to Norway, Faeroes, and south-west Iceland. Represented in Mediterranean by a form which may be identical and whose scientific names have included *Gobius reticulatus* Valenciennes and *G. affinis* Sanzo (not Kolombatovic).

Fins Pelvic anterior membrane and disc complete; edge of anterior membrane lacking villi and lobes; no free pectoral rays; 1st dorsal fin elongate in males. *Nostril* No process on rim of front nostril. *Scales* None on nape; 25-30 in lateral series. *Fin rays* 1D VI; 2D I, 8-9; A I, 7-8.

Coloration of body a coarse dark reticulate pattern flecked with rusty dots and interrupted by paler dorsal saddles opposite large single black spots on lateral midline. Length over 6 cm.

Not usually found close inshore, typically in over 10 m. and known even from deep water (330 m.) beyond the continental shelf. Found on coarse shell to offshore muddy grounds. Breeds from March-August (British Isles).

Four-spotted Goby p. 263
Deltentosteus quadrimaculatus (Valenciennes)

Mediterranean.

Fins Pelvic anterior membrane and disc complete, the anterior membrane with neither villi nor lobes; no free pectoral fin rays; 1st dorsal fin rays elongate, reaching to the front half of the second dorsal when depressed. *Scales* Nape and breast completely scaled; 33-35 scales in lateral series. *Fin rays* 1D VI; 2D I, 9; A I, 8.

Coloration fawn with reticulate pattern and four large black spots along lateral midline, opposite paler saddles across back; first dorsal fin with conspicuous black spot at posterior end and black streak along anterior edge, including elongate second ray; dusky pelvic and anal fins.
Length 8 cm.

Inshore on sand or muddy sand, to 90 m. Breeding season March to August (Western Mediterranean).

Deltentosteus colonianus (Valenciennes) p. 263
= *Gobius lichtensteinii* Steindachner.

Mediterranean.

Similar to the Four-spotted Goby, *D. quadrimaculatus*, but longer jaw (angle of mouth below rear half of eye) and more rays in 2D (I, 10-11) and A (I-10-11). Biology unknown in detail.

Diminutive Goby p. 264
Lebetus scorpioides (Collett)
= *Lebetus orca* (Collett)

Eastern Atlantic, from northern Bay of Biscay to Hennefjord (Norway), the Faeroes, and south-west Iceland.

Fins No anterior membrane to pelvic disc which is emarginate posteriorly; none of the pectoral rays are free; 1st dorsal fin elongate, especially voluminous in males, reaching when depressed to at least the middle of

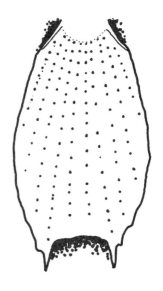

the second dorsal fin. *Nostrils* Front and rear nostrils tubular but without processes. *Scales* No scales on nape; 25-29 (usually 26-28) in lateral series. *Fin rays* 1D VI-VII (usually VI); 2D 9-11 (usually 9-10; A I, 7-8.

Coloration with vertical bars on body and broad, well demarcated, pale band across tail stalk; males yellowish to dusky grey, underside of head and breast reddish orange, first dorsal fin dusky yellow edged with white, second dorsal fin with intense black edge and oblique yellow and white bands; females pale brown, with bars purplish brown, and first dorsal fin displaying oblique yellow to orange-red bands with black spot at rear end, while second dorsal fin has narrow dark edge and thin oblique orange-red bands.
Length 3.9 cm.

Found at a wide range of depths, 2-375 m., but typically a more offshore species on coarse grounds, especially in the presence of calcareous (coralline) algae. Breeds from March-August (Isle of Man).

Coloration reddish to olive brown, with dark reticulate pattern and several pale saddle-markings on nape and back; many small alternate dark and pale bluish marks along lateral midline; dorsal fins banded with red; a large black spot partly edged with yellow at base of tail fin. Males with another large intense black spot in lateral midline below first dorsal fin.
Length over 6 cm.

Inshore, living above sea bed, in groups about weed-grown structures and over beds of oarweed or eel-grass, down to 16 m. Also in shore-pools among weed.

Sand Goby   p. 263 **pl. 118**
Pomatoschistus minutus (Pallas)

Mediterranean and Black Sea; Eastern Atlantic, from Spain to Tromso, Norway.

Fins Pelvic disc and membrane complete, the latter edged with minute villi; no free

Two-spotted Goby   p. 264 **pl. 107, 108**
Gobiusculus flavescens (Fabricius)

Eastern Atlantic, from north-western Spain to Vesteralen (Norway), the Faeroes, and western Baltic. Recorded once from Mediterranean (Palermo, Sicily).

Body somewhat laterally compressed. *Eyes* lateral with a wide space between the orbits. *Fins* Pelvic anterior membrane and disc complete, without villi along the edge of former; no free pectoral rays. *Nostrils* Front nostril simple and tubular, with no process on rim. *Scales* None on nape; 35-40 in lateral series. *Fin rays* 1D VII-VIII (usually VII); 2D I, 9-10; A I, 9-10.

pectoral rays; tail fin rounded. *Nostril* Front nostril rim without flap. *Scales* on rear part of nape; 58-70 scales in lateral series. *Fin rays* 1D VI-VII (usually VI); 2D I, 10-12; A I, 10-12.

Coloration sandy with ferruginous specks, fine reticulation, and small dorsal saddles. Males with conspicuous dark blue spot on first dorsal fin, in distal part of membrane between 1D V and 1D VI, dorsal fins with

reddish brown bands, four main vertical lateral bars across flanks, and dusky pelvics, although breast is unpigmented in Atlantic populations.
Length 9.5 cm.

Inshore sandy grounds, 0-20 m.; juveniles in estuaries. Breeds from March-July (Sweden; Brittany).

Pomatoschistus norvegicus (Collett) is an offshore form, usually regarded as distinct from *P. minutus* and occurring from 30-280 m. on coarse shell to mud, in both the Mediterranean and eastern Atlantic as far north as the Lofotens. In comparison with the preceding species, *P. norvegicus* is smaller (less than 6.5 cm.), paler, and males have more numerous but thinner vertical striations across flanks. 1D VI (VI-VII), 2D I, 8-10; A I, 8-10; scales in lateral series 55-58.

p. 263

Marbled Goby p. 263
Pomatoschistus marmoratus (Risso)
= *P. microps leopardinus* (Nordmann)

Mediterranean and Black Sea; also in Eastern Atlantic (south-west Portugal).

Fins Pelvic disc and anterior membrane complete, the latter with villi along the free edge; no free rays in pectoral fin; tail fin rounded. *Nostrils* Front nostril without flap. *Scales* Absent from nape but present on rear part of breast; 40-46 scales in lateral series. *Fin rays* 1D VI; 2D I, 8-9; A I, 8-9.

Coloration sandy, with dark reticulation and saddles; males with usually four vertical bars across flank, black spots in distal part of first dorsal fin (1D V/VI), dark reddish brown bands across median fins, and dusky pelvic fins and breast; females with intense black spot under chin.
Length nearly 6.5 cm.

Inshore, in sandy shallows; breeding season in spring and summer.

Painted Goby p. 263 **pl. 117**
Pomatoschistus pictus (Malm)

Mediterranean (Adriatic Sea); Eastern Atlantic, from western Channel to Tromsø, Norway. Belt Seas, but not Baltic proper.

Fins Pectoral disc and anterior membrane complete, the latter without villi; no free pectoral rays; tail fin rounded. *Nostrils* Front nostril without flap. *Scales* None on breast or nape; 36-43 scales in lateral series. *Fin rays* 1D VI; 2D I, 8-9; A I, 8-9.

Coloration fawn with coarse dark reticulation and large pale dorsal saddles reaching lateral midline on each side; four 'double' spots along lateral midline. In both sexes, dorsal fins with rows of large black spots, surmounted by rosy oblique banding.
Length 5.7 cm.

Inshore, low-water to 50 m., on gravel or sand, etc.; sometimes found in shore pools. Breeds from early April to July (Isle of Man).

Common Goby p. 264
Pomatoschistus microps (Krøyer)

Western Mediterranean; Eastern Atlantic, from southern Spain to Trondheim, Norway, and the Baltic.

Fins Pelvic disc complete; pelvic anterior membrane without villi; no free pectoral rays; tail fin rounded. *Nostrils* Front nostril without flap. *Scales* None on nape or as far back to the end of the first dorsal fin; no scales on breast; 39-52 scales in lateral series. *Fin rays* 1D VI-VII (usually VI); 2D I, 8-10; A I, 8-10.

Coloration greyish to fawn, with saddles and coarse reticulation; males with dark spot on proximal part of membrane between first dorsal rays V and VI, dorsal fins banded with reddish brown, dark pelvic disc, breast and underside of head (tinged

with orange), and up to about ten lateral bars across flanks.
Length 6.4 mm.

Inshore, occurring in estuaries as well as salt-marsh and high shore pools. Breeds from April-August (Isle of Man, Baltic Sea).

Tortonese's Goby p. 264
Pomatoschistus tortonesei Miller

Western Mediterranean (recorded only from Marsala, Sicily).

Fins Pelvic disc and anterior membrane complete, but no villi along the edge of the membrane; no free pectoral rays; tail fin rounded. *Nostril* Front nostril without process on rim. *Scales* None on nape and breast, 29-35 (usually 31-33) scales in lateral series. *Fin rays* 1D VI; 2D I, 7; A I, 7.

Coloration in life not recorded; both sexes with lateral bars, about 4-6 in males; fewer bars in females and that beneath origin of first dorsal fin not reaching below lateral midline.
Length about 3 cm.

Recorded from brackish water. Biology unknown. The Lagoon Goby, *Knipowitschia panizzae* (Verga), is another small species (about 3.8 cm.), found in brackish and fresh waters of the Adriatic basin; it differs from *P. microps* and *P. tortonesei* in lack of scales on back until middle or end of second dorsal fin, very prominent dorsal saddles, dark oblique band across first dorsal fin, and black blotch under chin in females. Scales in lateral scales 33-39; fin-rays 1D V-VI; 2D I, 7-9; A I, 7-8.

Kner's Goby p. 263
Pomatoschistus knerii (Steindachner)

Western Mediterranean and Adriatic.

Fins Pelvic disc and membrane complete, membrane without villi along free edge; tail fin square-cut but with rear corners slightly extended (truncate/emarginate).
Nostrils Front nostril without a flap. *Scales* None on nape or as far back as second dorsal fin; no scales on breast, 38-46 (usually 41-44) scales in lateral series. *Fin rays* 1D VI-VII (usually VI); 2D 9-11 (usually 9-10); A I, 8-10 (usually 9-10).

Coloration reddish orange with saddles, vertical bars on flanks and large spot at base of caudal fin in both sexes; in males, bars darker and more numerous, and dark blotch in posterior corner of first dorsal fin.
Lengh about 4 cm.

Inshore; precise habitat unrecorded; but may live above sea-bed.

Quagga Goby p. 263
Pomatoschistus quagga (Heckel)

Western Mediterranean and Adriatic.

Fins Pelvic disc and anterior membrane complete, the anterior membrane without villi; no free pectoral rays; tail fin may have a straight rear margin (truncate) (as in *P. knerii*); *Nostrils* Front nostrils without flap. *Scales* None on nape and at least as far back as the end of the 1st dorsal fin; 35-40 scales in the lateral series. *Fin rays* 1D VI-VII (usually VI); 2D I, 9; A I, 8-9 (usually 9).

Coloration with coarse reticulation between saddles, lateral bars in both sexes (four in males, three in females, with none under first dorsal in latter sex).
Length 4-5 cm.

Probably inshore, but biology largely unknown.

Transparent Goby p. 266
Aphia minuta (Risso)

Mediterranean and Black Sea; Eastern Atlantic, to Trondheim, Norway, and western Baltic.

Body Laterally compressed. *Eyes* Lateral with the orbits well separated. *Fins* Pelvic membrane and disc complete, larger in males; no free pectoral rays. *Teeth* adult males with large canine-like teeth. *Nostrils* Rim of front nostril without process. *Scales* Cycloid, those on body easily lost; scales do not extend onto nape; 24-25 scales in lateral series. *Fin rays* 1D V; 2D I, 11-13; A I, 12-14.

Body transparent; minute dots of pigment along bases of median fins and on head.
Length 5.1 cm. (males), 4.6 cm. (females).

Pelagic, inshore and estuarine, surface to 66 m., over sand, mud, eel-grass, etc.; feeds on plankton. Breeding season from June to August (Oslofjord). Adults subsequently die after merely one year of life.

Ferrer's Goby p. 266
Pseudaphya ferreri (O. De Buen & Fage)

Western Mediterranean and Adriatic.

Body Compressed. *Fins* Pelvic disc and membrane complete; no free pectoral rays; tail fin rounded. *Nostrils* Front nostril without flap. *Teeth* Similar in both sexes. *Scales* Ctenoid, absent from nape; 25-26 scales in lateral series. *Fin rays* 1D V; 2D I, 8; A I, 10.

Body transparent with rosy dots; reddish line along bases of median fins; large triangular or rhomboidal dark reddish spot at base of caudal fin.
Length 2.9 cm.

Inshore, on sandy ground, especially from December to February.

Another little-known western Mediterranean species, *P. pelagica* F. De Buen, differs from *P. ferreri* in the head patterns of sensory papillae.

Crystal Goby p. 266
Crystallogobius linearis (Von Düben)
= *Crystallogobius nilssonii* Von Düben & Koren

Mediterranean; Eastern Atlantic from Gibraltar to Lofoten Islands (Norway) and Faeroes.

Body Laterally compressed. *Eyes* Lateral with a wide space between orbits. *Fins* Pelvic disc a deep funnel-like structure in males; both pelvic disc and also first dorsal fin vestigial or lacking in females; no free pectoral rays. *Jaws & Teeth* Lower jaw of male with prominent canine teeth; jaws in female smaller and toothless. *Scales* Absent. *Fin rays* 1D II (males), O (females); 2D I, 18-19; A I, 20.

Coloration transparent, with dots of pigment on chin and along bases of median fins.
Length 4.7 cm. (males), 3.9 cm. (females).

Pelagic, but more offshore in distribution than Transparent goby *Aphia minuta*, in depths of even 400 m., over bottoms of dead shells, sand, or mud. Males live on sea-bed during breeding season (May to August), guarding eggs deposited in empty tubes of large worms such as *Chaetopterus*. Feeds on plankton.

Sub-order: S C O R P A E N O I D E I

Family:

TRIGLIDAE

The **Gurnards** have large armoured heads. The most characteristic feature is in the pectoral fins which have a few rays free from the fin membrane and which have taste buds scattered over the surface. The Gurnards delicately probe the bottom with their feelers in their search for food. Very superficially the Gurnards resemble the Red Mullet but differ in having pectoral feelers and not barbels.

Piper   p. 280
Trigla lyra L.

Mediterranean, east Atlantic from northern Scotland to Senegal.

Body tapers steeply from behind the head to the tail; *head* armoured, *profile* steep and concave; *spines*, there are two bony toothed lobes protruding in front of the snout. There is a long pointed spine situated immediately above the pectoral fin that is about half the length of the fin. On each side of the dorsal fin is a row of 24-25 short robust spines; *scales*, small with the free edge finely toothed; *lateral line* distinct and smooth, *fins* 2 dorsal fins, one anal fin similar to and opposite the 2nd dorsal fin. Pectoral fin long, extending to the 5th ray of 2nd dorsal and anal fin. The 3 lowest rays of the pectoral fin are free and modified into sensory feelers.
1D IX-X; 2D 16-17; A 16-17.
Up to 40 cm., occasionally 60 cm.
Red in colour, darker on the back, lighter on the flanks and silvery underneath. Fins reddish, the dorsal and pectoral fins often spotted with blue.

This species is usually found in deep water from 100-700 m.
Occasionally, however, it may penetrate into shallower water and has been caught at around 50 m. Feeds on small crustacea, echinoderms etc.

**Tub Gurnard; Saphirine Gurnard;
Yellow Gurnard**   p. 280 **pl. 127**
Trigla lucerna L.
= *Trigla hirundo* L.
= *Trigla corax* Bonaparte

Mediterranean, Black Sea, east Atlantic from Norway to Senegal, English Channel, North Sea.

Body tapers gradually from behind the head to the tail; *head* armoured; *profile* only slightly concave and less steep than in the Piper (*Trigla lyra*); *spines*, there are two short lobes either side of the snout. The spine above the pectoral fin is short. On each side of the dorsal fin there is a row of 24-25 small spines; *scales*, small on body, 70 larger scales along the lateral line; *lateral line* distinct and smooth; *fins*, 2 dorsal fins, 1 anal fin opposite to but rather shorter than the 2nd dorsal fin. Pectoral fin large and extends to the 3rd or 4th ray of the anal fin; the 3 lowest rays are free and modified into sensory feelers.
1D VIII-X; 2D 15-18; A 14-17.
Up to 65 cm., usually about 30 cm.
Brilliantly coloured and very variable. The back may be red, reddish-yellow, or brown or yellowish with brown or even greenish blotches. The flanks are reddish or yellowish shading to pink or white on the undersurface. The head is dark red. Pectoral fins have bright blue rims, blue spots and a blue-black blotch on the undersurface with light blue spots on the upper surface.

Piper *Trigla lyra* p. 279

Large toothed lobes extending either side of the snout;
pectoral fins reddish;
spines over pectoral fin long.
60 cm.

Tub Gurnard *Trigla lucerna* p. 279

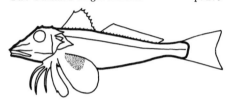

Spine over pectoral fins short;
pectoral fins blue with blue-black blotch on undersurface and light-blue spots on upper-surface.
65 cm.

Streaked Gurnard *Trigloporus lastoviza* p. 281

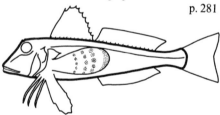

Skin arranged in distinct oblique stripes along the body;
head and body red, pectoral fin with rows of blue spots.
40 cm.

Grey Gurnard *Eutrigla gurnardus* p. 281

Greyish with white spots, or reddish, with a black spot on the dorsal fin;
pectoral fin does not reach anal fin.
50 cm. (Mediterranean 30 cm.)

Red Gurnard *Aspitrigla cuculus* p. 282

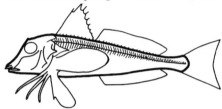

Large deep bony scales along the lateral line;
colour red.
30 cm.

Long-Finned Gurnard *Aspitrigla obscura* p. 282

2nd ray of 1st dorsal fin elongate;
silvery lateral line.
35 cm.

Armed Gurnard *Peristedion cataphractum* p. 282

Body and head covered with large bony plates;
large branched barbel on lower jaw;
25 cm.

This species ranges in depth from 5-300 m., and is found on sand, mud or gravel bottoms. Spawning occurs during early summer in the Atlantic and Channel and during winter in the Mediterranean. The young fish are frequently encountered near coasts and particularly in or around estuaries and even in fresh water where they rest on the bottom with their pectoral fins spread out. The adult fish feed mainly on crustacea. Like many of the Triglidae this fish is known to emit noises.

Streaked Gurnard   p. 280 **pl. 126**
Trigloporus lastoviza (Brünnich)
= *Trigla lineata* Pennant
= *Trigla adriatica* Gmelin

Mediterranean, east Atlantic north to Scotland, south to Canaries, English Channel.

Body tapers from behind the head to the tail; *head* covered with bony plates; *profile* steep and slightly concave, *snout* blunt, without projecting spines; *spines* there are small spines above the eyes. The spine above the pectoral fin is short but has a very wide base; there are 25 spines along the body either side of the dorsal fins; *skin*, arranged in distinct oblique stripes along the body; *fins*, 2 dorsal fins, anal opposite and similar to the 2nd dorsal. Pectoral fins reach the 3rd-7th anal fin rays. The lowest 3 rays of the pectoral fin are modified into sensory feelers.
1D IX-XI; 2D 16-17; A 14-16.
Back red sometimes with darker blotches, sides paler and belly white. Dorsal fins are pinkish and the anal fin pinkish at the base and yellowish at the tip. Pectoral fins reddish with dark blue spots arranged in rows.
Up to 40 cm.

Most frequently found on muddy or sandy areas near rocks from 45 m. Spawning occurs in the summer. Adult fish feed exclusively on small crustacea.

Grey Gurnard   p. 280 **pl. 125**
Eutrigla gurnardus (L.)
= *Trigla milvus* Lacépède

East Atlantic from Norway and Iceland to Senegal, Mediterranean, west Baltic, North Sea, English Channel, also Black Sea.

Body tapers from behind the head to the tail; *head* with bony plates; *profile* sloping, almost straight; *snout* with a lobe projecting on either side each with 3-4 small spines; *spines*, the spine above the pectoral fin is short. 27 bony spines are present either side of the dorsal fins; *lateral line* 70-77 scales each with a central ridge with a point at the end and toothed free edge; *fins*, 2 dorsal fins, anal fin opposite to but shorter than the 2nd dorsal fin. Pectoral fin does not reach the anal fin. The three lowest rays of the pectoral fin are modified into sensory feelers.
1D VII-IX; 2D 18-19; A 17-20.
Variable in colour. The Atlantic form is usually grey, greyish-red or greyish-brown with numerous whitish-yellow spots scattered over the back and sides and the lateral line is also yellowish-white. The 1st dorsal fin has a black patch. The Mediterranean form is usually red or reddish-brown but also has a black patch on the 1st dorsal fin.
Up to 50 cm., the Mediterranean form reaches 30 cm.

Very common in the Atlantic where they are found in groups on muddy or sandy bottoms or on sandy areas amongst rocks. They inhabit all depths down to 200 m. During the summer there is a migration into shallow water when they may even penetrate into estuaries. They feed predominantly on crustacea but also on small fish. Spawning occurs from January to September, during the later months in the northern ranges and in the early months in the southern ranges. Like the other Gurnards this species is able to emit noises.

Red Gurnard p. 280 **pl. 124**
Aspitrigla cuculus (L.)
= *Trigla pini* Bloch

East Atlantic north to Norway south to Mauritania; southern North Sea, English Channel, Mediterranean.

Body tapers from behind the head to the tail; *head* covered with bony plates; *profile* fairly steep and somewhat concave; *scales* 65-70 along the lateral line, gradually decreasing in size towards the tail. These scales are very deep and bony but not toothed or spined; *spines*, 3-4 small spines project either side of the snout. The spines above the pectoral fin are strong but not large. There are 26-28 spines either side of the dorsal fins; *fins*, 2 dorsal fins, anal fin similar to and opposite the 2nd dorsal fin. The pectoral fins extend to about the 3rd ray of the anal fin.
1D VIII-X; 2D 17-18; A 16-18.
Head, back and flanks reddish, undersurface white. Pectoral fins may be greyish, pinkish or yellowish. The tail fin has a pale coloured base and a dark rear edge.
Up to 45 cm., commonly 20-30 cm.

In Atlantic waters this fish is very common and ranges in depth from 5-250 m. In the Mediterranean they are normally caught between 100-250 m., but sometimes in much more shallow water near coasts (30 m.) on sandy, muddy, gravel, shell or rocky areas with sandy patches. Spawning occurs during the spring and summer. The adults feed mainly on crustacea but also on fish.

Long-Finned Gurnard; Shining Gurnard
Aspitrigla obscura (L.) p. 280

Mediterranean; East Atlantic north to English Channel.

Body tapers from behind the head to the tail; *head* covered with bony plates; *profile*, more or less straight; *spines*, there is a bony lobe with 1 distinct spine present on either side of the snout. The spine above the pec-

toral fin is very short. On each side of the dorsal fin is a row of 27-28 bony plates with toothed margins; *scales*, very small on body; large, 68-70, along the lateral line and untoothed; *fins*, 2 dorsal fins. The 2nd ray of the 1st dorsal fin is elongate. The 2nd dorsal is similar to the anal fin. The pectoral fin does not extend beyond the 3rd anal fin ray and the 3 lowest rays are modified into sensory feelers.
1D IX-X; 2D 16-18; A 14-18.
The colour is very variable. The back is commonly greyish pink shading to pearly pink underneath. The lateral line is a silvery pink. Pectoral fins are bluish, dorsal fins and tail fin reddish. Pelvic fins and anal fin reddish sometimes with yellowish or white edges.
Up to 35 cm.

Found down to 50 m. on rocky and sandy bottoms. Little is known about the habits of these fish except that they feed mainly on small crustacea. They are believed to be bioluminescent.

Armed Gurnard; Mailed Gurnard p. 280
Peristedion cataphractum (L.)

Mediterranean; East Atlantic north to Gascony, south to Senegal.

Body long and slender, tapers from behind the head to the tail; *head* armoured; *snout*, there are 2 long, flat bony lobes which extend forward on either side and leave a deep concavity in the centre; *mouth* situated underneath; *teeth* absent; barbels, the lower jaw has a number of small barbels and 1 pair of larger, much branched barbels; *scales*, body completely covered in large bony scales each with a central ridge 27-29 along the flanks; *fins*, 2 dorsal fins. The rays of the 1st dorsal fin are all elongate and continue beyond the fin membrane. The anal fin is opposite but shorter than the 2nd dorsal fin. The pectoral fins have the lowest 2 rays modified into sensory feelers.
1D VII-VIII; 2D 18-19; A 18-21.
Red in colour, darker on the back and head

and lighter on the belly.
Up to 40 cm., usually 25 cm.

Usually found in deep water from 50-500 m.

and most frequently encountered on mud where they feed on the small animals living in it.

Family:

DACTYLOPTERIDAE

The **Flying Gurnards** are separated from the Gurnards (Triglidae) by the primitive characteristics of their head bones.

Flying Gurnard   pl. 128, 129
Dactylopterus volitans

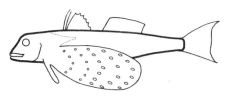

Very large pectoral fins with bright blue spots and stripes.
50 cm.

Mediterranean; east Atlantic north to Brittany, south to Angola.

Head large with a corselet type structure extending onto the body; *snout* rounded; *eyes* large; *fins*, the pectoral fins are very large and make this fish unmistakable. They are divided into two sections, the front section

is short and consists of 6 rays and is attached by the base only to the second, much larger, section. The 1st and 2nd rays of the 1st dorsal fin are separate.
1 D, II + IV; 2D I + 8; A 6.
Head, back and flanks greyish or brownish often with darker spots and blotches, and lighter spots. Undersurface pinkish-white. The pectoral fins are brown with bright blue spots and stripes arranged in a regular pattern.
Up to 50 cm.

Live on sandy or muddy bottoms, usually around 10-30 m. They feed mainly on small bottom living animals predominantly crustacea. Spawning occurs during summer. The very large pectoral fins are usually folded back when the fish is resting on the bottom but when disturbed the fins open and the fish 'fly' through the water.

Family:

SCORPAENIDAE

Robust-bodied fishes with the head armoured with spines. There is one dorsal fin but the spiny part is divided from the soft-rayed parts by a pronounced dip. There are no free rays on the pectoral fin. In preserved specimens the spines on the head are useful for identification but in life they are frequently hidden by tissue and are not very useful for field identifications.

The **Scorpion Fishes** are perhaps the most dangerous animal that most Mediterranean swimmers will encounter. At the base of the spiny rays lie venom glands and the poison is directed to the tip of the spine along a canal.

The treatment is to plunge the affected part into very hot water until the pain decreases. Wounds of this nature often go septic and should be thoroughly cleaned and sterilized.

Scorpion Fish   pl. 196
Scorpaena porcus L.

Plume-like appendages over eye; no appendages on chin.
25 cm.

Mediterranean and eastern Atlantic from the Canary Islands to Biscay. Very rarely as far north as the English Channel.

Head heavily armoured with spines; *eye* large and oval; *appendages*, there is a well developed plume-like appendage above each eye and two smaller ones on the front nostrils. There are no appendages on the chin; *scales*, small, at least 55 along the lateral line.
D XII, 9-11; A III, 5-6.
Usually brown or reddish-brown with irregular spots and bands which make it extremely difficult to see in the water.
Up to 25 cm.

During the daytime it usually lies motionless amongst rocks in shallow water. It is extremely difficult to see and a swimmer may inadvertently touch the poisonous spines of the dorsal fin and gill cover (see above for treatment). When disturbed they will swim off at high speed but soon settle again. They are carnivorous and are believed to feed at night. Breed in late spring and summer. The eggs are laid and embedded in a transparent mucous lump.

Scorpion Fish   pl. 122, 123, 197
Scorpaena scrofa L.

Numerous appendages on chin.
40 cm.

Mediterranean and eastern Atlantic from Senegal to Biscay.

Body robust, head armoured with spines; *eye* oval, smaller than in *S. notata* and *S. porcus*; *appendages*, less well developed than in *S. porcus* except on the chin where they are numerous; *scales* rather large (35-40 along the lateral line).
D XII, 9-10; A III, 5-6.
Colour extremely variable, usually reddish-brown with dark and light mottling. In natural light underwater the red colour is not visible and the fish closely resembles the background of rock and weed where it lies. Flash photography as in our photograph shows up the red colour and the fish no longer looks camouflaged. There is often a black spot in the middle of the dorsal fin.
Up to 40 cm., rarely 50 cm.
Like the other Scorpion fishes, they typically rest on the bottom in the daytime. The young fish may be found in shallow water but the adults normally stay between 20 and 200 m. Usually on rocky ground. Their general biology is similar to *S. porcus* but live in rather deeper water. For treatment of stings see above.

Scorpion Fish
Scorpaena notata Rafinesque
= *Scorpaena ustulata* Lowe

No prominent tentacle over eye;
no appendage on chin;
black spot on dorsal fin.

Mediterranean and Eastern Atlantic from
Senegal to the Gulf of Gascony.

Body robust, the head is heavily armoured
with spines; *eye* large and nearly round; *appendages*, there are no very well developed
appendages. There are small ones on the
nostrils, a very short one above each eye and
none on the chin; *scales* are relatively large
(36-40 along the lateral line).
D XII, 9-10; A III, 5.

Brick red with variable mottling. Frequently there are dark spots on the dorsal, anal
and tail fins. A constant feature of this fish
is a dark blotch between the 8th and 10th
dorsal spines. Note that a similar spot is
frequent on *Scorpaena scrofa* as well.

Family:

COTTIDAE

The head is wide, the gill cover has at least one spine; skin scaleless but often with hard
plates of varying size embedded in it.

Father Lasher; Short-Spined Sea Scorpion;
Bull Rout   p. 286 **pl. 198**
Myoxocephalus scorpius (L.)
= *Cottus scorpius* (L.)

North Atlantic south to Biscay; Iceland;
Greenland; North Sea; English Channel.

Head wide, flattened from above and with
bony crests; *jaws*, no fleshy barbels at the
corners of the jaws; *spines*, 2 spines present
on the front gill cover, the upper longer than
the lower but less than the eye diameter;
skin smooth, without scales but with bony
plates embedded in it and spines either side
of the lateral line; *lateral line* smooth; *gill
cover*, the gill covers are joined by a skin
membrane which forms a flap across the
body of the fish; *fins*, 2 dorsal fins, 1 anal
fin. Pectoral fin large, pelvic fin with 3 rays.
1D VII-XI; 2D 14-17; A 10-14.
This fish is very variable in colour. Frequently greyish or brownish with darker
blotches on the head and back and with
lighter spots on the flanks. Belly yellowish
in the male and orangey in the female. The
fins are light with darker spots and bars.
During the breeding season the males are
coppery with lighter spots.
Up to 30 cm.

In the northern ranges of its distribution this
fish is very common and found near the
shore line. In the more southern water they
are found rather deeper from 2-60 m. They
inhabit a variety of bottoms including mud,
sand, shell, gravel and seaweed and also
penetrate into estuaries. Adult fish feed
mainly on crustacea, and fish. Spawning
takes place during winter. The eggs are laid
in small masses near rocks and seaweed.

Long-Spined Sea Scorpion   p. 286 **pl. 121**
Taurulus bubalis (Euphrasen)
= *Myoxocephalus bubalis* (Euphrasen)
= *Cottus bubalis* Euphrasen

285

Father Lasher *Myoxocephalus scorpius*
p. 285

2 spines on front gill cover;
no fleshy barbel at corner of jaws;
smooth lateral line.
30 cm.

Long-Spined Sea Scorpion p. 285
Taurulus bubalis

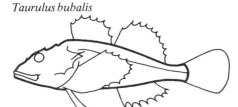

3 or 4 spines on front gill cover, uppermost
longer than eye diameter;
fleshy barbel at corner of jaws;
spiny lateral line.
Males 12 cm.
Females 25 cm.

Norway Bullhead *Taurulus lilljeborgi*

4 spines on front gill cover;
small barbel at corner of jaws;
row of spines above the lateral line;
northern distribution.
60 cm.

Four-Horned Sculpin p. 287
Onocottus quadricornis

4 bony lumps on head.
20 cm.

East Atlantic south to Portugal, North Sea,
English Channel, rarely Mediterranean.

Head large with bony crests; *jaws* with one
or two fleshy barbels at their corners; *spines*,
there are 3 or 4 spines present on the front
gill cover, the uppermost is longest, more
than the eye diameter, and pointed; *skin*
smooth without scales; *lateral line* consists
of a row of 32-35 spiny plates; *gill cover*,
the skin membrane extending from the gill
covers is joined to the body of the fish and
does not form a flap; *fins* 2 dorsal fins, 1
anal fin, pectoral fin large, pelvic fin with 3
rays.
1D VII-IX; 2D 10-14; A 8-10.
Variable in colour but frequently brown
with greenish and light blotches. Undersurface yellowish or silvery, particularly in the
breeding male. They may sometimes be reddish in colour when at first glance they may
be mistaken for a Scorpion fish.
Males 8-12 cm., females 10-25 cm.

Very common in shallow water and rarely
found below 30 m. They live amongst rocks
and particularly in areas where there is plenty of seaweed or sea grass. Their diet is very
varied and includes small crustacea, worms
and small fish which they catch by making
short forays from their lair. Breeding occurs
during spring and the eggs are laid in a mass
amongst the rocks and frequently in small
crevices.

Norway Bullhead
Taurulus lilljeborgi Collett
= *Acanthocottus lilljeborgi* (Collett)

West coast of Scandanavia to West Iceland;
Irish Sea.

This species is very similar to *Myoxocephalus bubalis* but may be distinguished by a
row of bony spines above and parallel to the
lateral line; the pelvic fins which only have
2 rays and the maximum size of 60 cm.
There are 4 spines on the front gill cover but
the uppermost is only equal to the eye diameter.

Light brown or yellowish with 4 darker, greyish, transverse bands on body and 1 on head. During breeding male fish have a red band behind the head and red patches along the flanks.
Up to 60 cm.

Found from 20 m. on coarse shell and gravel bottoms. Breeding occurs during spring in its southern range and summer in the more northern latitudes. The eggs are laid amongst stones.

Four-Horned Sculpin   p. 286
Oncocottus quadricornis (L.)

All coasts of the northern hemisphere northwards of Denmark. Not around the British Isles. Baltic and freshwater lakes.

Body rather tadpole shaped; *head* large, flat

and wide, there are 4 conspicuous bony warts on the top of the head; *scales* absent, there is a row of granular bumps between the lateral line and the back; *gill cover* the front gill cover has 3-4 spines of which the uppermost is the longest, equalling the diameter of the orbit; *fins* 2 dorsal fins, the 1st is low, 2/3 height of the 2nd dorsal. D VII-IX, 13-16; A 13-17.
Greyish-brown above tinged with red on the gill covers, yellowish on the sides, whitish below.
20 cm., rarely up to 60 cm.

Able to tolerate salt, brackish and fresh water. Its distribution throughout the cold waters of the northern hemisphere and its presence in landlocked lakes suggest that it was once associated with glacial conditions and remained in place when the ice retreated.

Family:

AGONIDAE

A group of temperate and cold water bottom-living fishes. The body is completely covered with bony plates. There are two dorsal fins, the head is spiny and the mouth is underneath.

Hook-Nose; Pogge; Armed Bullhead
Agonos cataphractus (L.)

Armoured body;
small barbels under head.
15 cm.

East Atlantic north beyond Iceland and Norway, south to English Channel, North Sea, western Baltic.

Head very wide and triangular when seen from above; *body* very deep behind the head, curves steeply to a long slender rear portion. Completely covered with hard bony plates; *profile* slightly concave; *snout* with one pair of spines on either side; *bar-*

bels, there are a large number of short barbels present on the lower surface of the head; *fins*, 2 dorsal fins set close together. 1D V-VI; 2D 6-8; A 6-7.
Brown with 4-5 darker saddles along the back. Undersurface whitish. The pectoral fin may be orange.
Up to 15 cm.

A very common fish found from shallow water to below 500 m. on mud, sand or stones but rarely seen by divers perhaps because it lies buried in the sand. In the southern range of its distribution it is frequently encountered in estuaries during the winter. Spawning occurs during autumn and winter; the yellow eggs are laid in clumps, frequently amongst kelp (*Laminaria*) holdfasts. Its diet is very varied and includes small crustacea, echinoderms and other shellfish.

Family:

CYCLOPTERIDAE

The **Lump Sucker** and **Sea Snails** are distinguished by having no scales and their pelvic fin is modified into an efficient sucker by which the fish attaches itself to stones, algal hold-fasts etc.

Lump Sucker; Sea Hen   **pl. 130, 131**
Cyclopterus lumpus L.

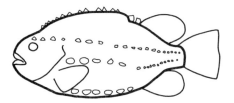

Pectoral fin forms a sucker;
bony plates on body.
60 cm.

Both sides of north Atlantic from the Arctic Sea south to Portugal.

Body massive, tall and rounded; *head* bony, armoured with bony plates of which there are four enlarged rows, one along the back, the second running backwards from the eye to the tail fin, the third from the angle of the mouth to the tail fin and the fourth from the pectoral fin base to the anal fin base; *fins* two dorsal fins in the young. In the adult the first dorsal becomes hidden by thick skin. The pelvic fins are united into a sucker. 1D 6-8; 2D I, 10; A I, 10.
Bluish or greyish; the males in the breeding season are orange or brick red below. The young are green.
Males up to 50 cm., females up to 60 cm.

A bottom-living fish found from very shallow water down to 300 m. and more. The well-developed sucker is used to attach the fish to rocks and other solid objects. In late winter and spring they come into shallow water sometimes even into the intertidal zone to breed. The eggs are pink and laid in a large mass. They take between one and two months to hatch during which time they are guarded with great tenacity by the male who fans water through the egg mass to keep them oxygenated. The female meanwhile has returned to deep water. The larvae live in the plankton.
The eggs are dyed, preserved and sold as 'caviar' substitute.

Sea Snail; Unctuous Sucker
Liparis liparis (L.)
= *Liparis vulgaris*

Tadpole-shaped body with sucker;
flaccid skin;
anal fin joined to tail fin.
12 cm.

North Atlantic from the Arctic Ocean to English Channel, sometimes in baltic.

Body rounded in front, laterally compressed behind; *head* broad and blunt-snouted; *sucker*, there is a well-marked round sucker set forward on the belly; *skin* loose and slimy extending over the fins, there are no scales; *fins*, the pelvic fin is modified into a sucker; pectoral fins rounded, nearly meeting under the head. The last ray of the anal fin is joined to the tail fin by a membrane. D 33-36; A 27-29.
Variable in colour usually of some drab shade that matches the bottom. Sometimes with indistinct streaks and lines.
Up to 12 cm., sometimes 18 cm.

Usually in shallow water, but its range extends from just beneath the surface to 100 m. amongst stones and in estuaries. It breeds in winter in shallow water usually near the mouths of rivers. The eggs are laid in clumps amongst hydroids or amongst fine algae. It feeds on small animals, chiefly crustacea.

Tadpole-shaped body with sucker;
flaccid skin;
anal fin not joined to tail fin.
Up to 9 cm.

Montagu's Sea Snail
Liparis montagui (Donovan)

Resembles the Sea Snail (*Liparis liparis*) except that the anal fin is not joined by a membrane to the tail fin. D 28-30; A 22-25. Up to 9 cm.

Lives in shallower water than *Liparis liparis*, it extends downwards from mid-tide level to 30 m. or more. Breeds in spring and early summer.

Family:

GASTEROSTEIDAE

The **Sticklebacks** are marine and freshwater fishes of cool northern waters. They have a series of spines along the back in front of the dorsal fin. The pelvic fin consists of one spine and 1-3 soft fin rays. The body is armoured with bony plates.

Stickleback; Three-Spined Stickleback
Gasterosteus aculeatus L.　　　　**pl. 200**

3 spines on back (may be folded back);
bony plates along flanks.
7 cm.

Fresh waters of Europe, North Africa and North America. Also in brackish and fully marine waters as far south as Northern Ireland and the North Sea.

Body oval and laterally compressed with a slender tail stalk. The flanks are armoured with a series of bony plates which are more numerous in salt-water specimens; *spines*

and *fins*, 3 spines on the back, the first two large, the third small situated just in front of the dorsal fin. These spines are usually folded back on the body and are thus not conspicuous. The pelvic fin consists of a single spine and a much shorter fin ray. The dorsal fin is longer than the anal. D III, 8-14; A I, 6-11.

Silver, olive or blue above, silver flanks and white belly. In the breeding season the males become a brilliant red below and the eye assumes an iridescent peacock blue.
Up to 7 cm.

The Stickleback can live both in fresh water and in fully saline coastal waters. It lives in most types of fresh water except for stagnant ponds. In the sea it probably prefers rocky areas with some weed. Spawning is chiefly from April to May. The male builds a nest on the bottom from strands of weed etc. and this he sticks together with a cement secreted from his kidneys. He herds the female into the nest to lay her eggs. These are guarded by the male and aerated by fanning with his pectoral fins. He also cleans the nest of unwanted fragments and drives off intruders. The nest is used for successive spawnings. The breeding behaviour of the Stickleback has been extensively studied since it involves complicated nuptial and threat displays. The red of the belly of the male evokes aggressive behaviour in other males. Their food consists of any aquatic animals small enough to eat.

The **nine-spined Stickleback** *Pungitius pungitius* is similar except for the extra spines along the back (8-10).

Primarily a fresh-water species it occasionally enters brackish water. **pl. 201**

Fifteen-Spined Stickleback **pl. 202**
Spinachia spinachia (L.)
= *Gasterosteus spinachia* L.

14-16 spines on back;
elongate body;
rounded tail fin.
20 cm.

North-east Atlantic from Biscay to north Norway. North Sea and Baltic as far north as Åland.

Body very long with a slender tail stalk; there is a raised ridge of about 40 plates in a line down the body; *spines* and *fins*. There are 14-16 short spines along the back. The pelvic fin consists of a single spine and a short fin ray. The tail is small and rounded; the dorsal and anal fins are almost opposite and set far back on the body. D XIV-XVI, 6-7; A I, 6-7.

The colour is brownish or greenish. In the breeding season the male becomes more blue in colour.

Up to 20 cm.

Typically a coastal marine species but can tolerate estuarine conditions. It is always found amongst weed sometimes even in rock pools and not deeper than 10 m. Spawning occurs in spring and early summer in very shallow water, sometimes even between tide marks. The nests which are built amongst sea weeds are woven from plant fibres and are about the size of a fist. It is cemented together with secretions from the kidney. There are 150-200 eggs which are laid inside the nest and are guarded by the male. The eggs hatch after about 20 days. The fish are said to live for only two years the larger fish being in their second year.

Order: HETEROSOMATA

This order contains the well-known **Flatfish.** The adult fishes lie on their side with both eyes on the uppermost side. The dorsal and anal fins are long and many-rayed. The upper 'eyed-side' is coloured usually in such a way as to camouflage the fish on the bottom; the under 'blind-side' is usually white.

The eggs are pelagic but hatch after a few days. The larvae are symmetrical like conventional fishes and swim in the upper layers of the sea. However, when they have reached a length of about 2 cm. one eye moves over the top of the head taking up station next to the other. At the same time the dorsal fin extends forward onto the head and the body becomes flattened. At about the time that the eye migrates, the fish comes to rest on the bottom to begin a fundamentally bottom-living existence.

Family:

Eyes on the left side of the head. The egg has a single oil-globule in the yolk. The front gill cover has a free edge. Pelvic fin not symmetrical. Mouth terminal.

Turbot   p. 292   **pl. 137**
Scophthalmus maximus (L.)
= *Psetta maxima* L.
= *Rhombus maximus* Cuvier

Eastern Atlantic, west Baltic, Mediterranean. There is a related species *Scophthalmus maeoticus* (Pallas) in the Black Sea.

Body very deep and rounded; *eyes* situated on the left side of the head and the lower eye is slightly in front of the upper; *mouth* large and situated on the left of the eyes; *profile*, there is a slight indentation at the origin of the dorsal fin; *skin*, there are no scales on the skin but the eyed-side has a number of bony tubercles scattered over the surface; *fins*, the dorsal fin commences just in front of the lower eye. The first few rays are slightly branched and the tip of the rays is free from the fin membrane. Neither the dorsal nor the anal fin extends under the tail. D 57-72; A 43-56.

The colour of the eyed-side is variable and depends greatly on the colour of the sea bed on which the fish is living. It ranges from light grey-brown to dark chocolate-brown with yellowish, light or dark brown, blackish or greenish spots. The fins are speckled including the tail fin which is densely covered with spots. Blind-side whitish.

Up to 80 cm., occasionally up to 100 cm. water. The younger, smaller, fish are commonly in the shallow water and the older, The Turbot is found on sandy, muddy, shell and gravel bottoms from very shallow water to below 80 m. also inhabits brackish water. The younger, smaller fish are commonly in the shallow water and the older, larger, specimens in the deeper water.

Breeding occurs during spring and summer. In northern European water the fish spawn over gravel bottoms at depths ranging from 10-80 m. At about 25 mm. when they are about 4-6 months old the young fish take up a bottom living existence. They are carnivorous and feed mainly on other fish. The turbot is of great commercial value, the majority of the fish being caught in the central North Sea.

Brill   p. 292   **pl. 203**
Scophthalmus rhombus (L.)
= *Rhombus laevis* Bonaparte
= *Bothus rhombus* Jordan & Gilbert

East Atlantic north to central Norway, North Sea, English Channel, western Baltic, Mediterranean.

Body oval; *eyes* situated on the left side of the head. The lower eye is slightly in front of the upper; *mouth* large and situated on the left of the eyes; *profile* smooth; *skin* covered with scales which are rather large on the eyed-side. There are no tubercles on the skin; *fins*, the first rays of the dorsal fin are partly free from the fin membrane and branched. Neither the dorsal nor the anal fin continues under the tail. D 73-83; A 56-62.

The colour of the eyed-side is very variable and ranges from sandy to grey brown to dark brown with many small irregular darker spots and other lighter spots scattered over the surface. The fins are lighter and the tail fin is slightly spotted. Blind-side white sometimes with occasional darker blotches.

Up to 70 cm.

The Brill may be found from very shallow water down to about 70 m. on sandy, muddy or gravel bottoms. They also penetrate

Turbot *Scophthalmus maximus*   p. 291

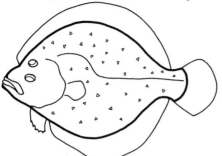

Skin without scales but with scattered tubercles;
tail fin heavily speckled.
80 cm.

Brill *Scophthalmus rhombus*   p. 291

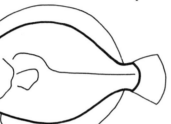

Skin with scales and no tubercles;
tail fin only slightly spotted;
first rays of the dorsal fin branched.
70 cm.

Topknot *Zeugopterus punctatus*   p. 293

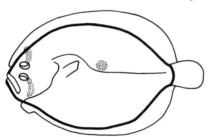

Pelvic fin attached to anal fin;
dorsal and anal fins form lobes under tail stalk;
dark spot behind curve of the lateral line;
dark bars extending from eyes.
25 cm.

Bloch's Topknot *Phrynorhombus regius*
p. 294

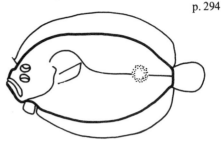

Dorsal and anal fins form lobes under tail stalk;
1st ray of dorsal fin elongate;
distinct notch in profile.
20 cm.

Norwegian Topknot   p. 294
Phrynorhombus norvegicus

Dorsal and anal fins form lobes under tail stalk;
profile smooth;
body slender.
Northern distribution.
12 cm.

Megrim *Lepidorhombus whiffiagonis*
p. 294

Slender body;
lower eye clearly in front of upper;
lower jaw prominent.
61 cm.

Lepidorhombus boscii p. 295

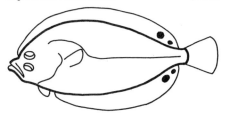

Body slender;
2 dark spots on dorsal and anal fins.
41 cm.

Wide-Eyed Flounder *Bothas podas* p. 295

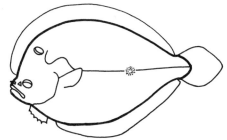

Eyes widely separated;
1 or 2 dark spots on lateral line towards tail.
Male fish with a spine in front of lower eye and another on snout.
20 cm.

Scaldfish *Arnoglossus laterna* p. 295

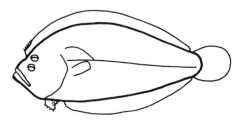

Scales very easily removed;
first rays of the dorsal fin not elongate but partly free from fin membrane.
D 84-98; A 63-75.
19 cm.

Arnoglossus thori p. 296

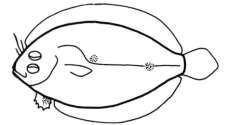

First rays of the dorsal fin not attached to fin membrane;
2nd ray of dorsal fin elongate.
18 cm.

Arnoglossus imperialis p. 296

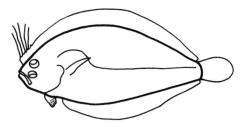

2nd-5th or 6th rays of the dorsal fin elongate, particularly in adult male.
25 cm.

into areas of brackish water. The larger, older, fish are usually found in the deeper water and the younger, smaller, specimens in the shallow water. Breeding occurs during spring and summer in the Atlantic, North Sea and English Channel and during February to March in the Mediterranean. The adults feed mainly on fish. They are edible but not of great commercial importance. Hybrids between the Brill and the Turbot are sometimes found.

Topknot p. 292 **pl. 208**
Zeugopterus punctatus (Bloch)

East Atlantic south to Biscay; English Channel; North Sea.

Body very deep, oval (height more than 1/2 length); *profile* smooth; *eyes* situated on the left side of the head one above the other; *mouth*, large situated on the left of the eyes; *scales* toothed on the eyed side and smooth on the blind side; *fins* dorsal fin commences very far forward on the snout. Both the dorsal and anal fin clearly continue under the tail stalk where they form distinct lobes. The pelvic fins are joined to the anal fin. D 88-102; A 67-76.
Eyed-side brown with darker blotches. There is a dark spot just behind the curve of the lateral line and dark bars extending from each eye.
Up to 25 cm.

Usually found from shallow water down to 37 m. Although fairly common they are extremely difficult to find as they have the habit of clinging to the bottoms and sides of large rocks where they are almost invisible. Spawning occurs in late winter.

Bloch's Topknot; Eckström's Topknot
Phrynorhombus regius (Bonnaterre) p. 292
= *Zeugopterus unimaculatus* Day   **pl. 136**

East Atlantic north to Shetland Isles, Mediterranean.

Body deep and oval; *profile* with a distinct notch in front of the upper eye; *eyes* on the left side of the head. They are large and situated one above the other or with the lower eye slightly in front of the upper. There is a bony ridge separating the two eyes; *mouth* large and situated on the left of the eyes; *scales* small with toothed free margins on the eyed side and less so on the blind side. 72-80 scales along the lateral line; *fins* the dorsal fin commences in front of the eyes and just behind the notch present in the profile. The 1st ray of the dorsal fin is longer than the others. Both the dorsal and the anal fins continue onto the blind side of the tail stalk. D 73-80; A 60-68.
Eyed-side reddish-brown with darker spots and blotches. There is one large dark spot often with a lighter centre situated on the

lateral line towards the tail. The dorsal and anal fins have dark spots which are more or less regularly arranged.
Up to 20 cm.

This species is not common in the Atlantic waters. It lives from 9-55 m. on a variety of bottoms both sandy and rocky. In the Mediterranean they have been caught from between 100-300 m. It is possible that these fish cling to the sides of rocks in a similar fashion to the Topknot (*Zeugopterus punctatus*). Spawning occurs in spring and summer.

Norwegian Topknot   p. 292
Phrynorhombus norvegicus (Günther)

East Atlantic south to English Channel; north to Iceland.

Body oval (depth much less than 1/2 body length); *profile* smooth; *eyes* situated on the left side of the head. They are very close together, one above the other; *jaws* situated on the left of the eyes. They extend just beyond the front edge of the eyes; *scales* large, finely toothed on the eyed side, less so on the blind side; *fins* dorsal fin starts in front of the eyes and both the dorsal and anal fins clearly continue under the tail stalk where they form distinct lobes. Pelvic fins separate from anal fins. D 76-84; A 58-68.
Eyed-side sandy brown with darker blotches. The dorsal and anal fins have irregular dark markings.
Up to 12 cm.

This species is the smallest and most northern of the Topknots. Very little is known about it habits except that they live in rocky areas and are usually found down to 50 m. sometimes down to 170 m. Spawning occurs in spring and early summer and the young fish take up a bottom living existence at about 13 mm.

Megrim; Sail-Fluke   p. 292
Lepidorhombus whiffiagonis (Walbaum)
= *Pleuronectes megastoma* Donovan

East Atlantic north to Iceland south to Morocco; North Sea; western English Channel; western Mediterranean.

Body long, slender, oval; *snout* rather long; *eyes* situated on the left side of the head. They are large, oval, close together and separated by a bony ridge. The lower eye is clearly situated in front of the upper; *jaws* on the left of the eyes. The lower jaw is longer than the upper and has a small bony spine at the tip; *scales* small, 95-109 along the lateral line; *fins* the dorsal fin starts immediately in front of the front eye. The dorsal and anal fins only just continue under the tail stalk. The longest rays of the dorsal and anal fins are found towards the rear of the fins. D 80-94; A 61-74.
The eyed-side is sandy brown with darker blotches. The eye iris is greenish yellow and the pupil is circled with bright yellow. Blind-side white.
Up to 61 cm.

Found on sandy bottoms from 10-600 m. but most commonly from 50 m. in the Atlantic and rather deeper in the Mediterranean. Only very rarely are they found in shallow water or near coasts. They feed on large quantities of many varieties of fish but also on crustacea. Breeding occurs in water between 50-200 m. deep from March to June. When the young fish reach about 19 mm. they take up a bottom living existence.

Lepidorhombus boscii (Risso)   p. 293

Eastern Atlantic north to Orkneys; Mediterranean.

This fish is very similar to *Lepidorhombus whiffiagonis* except that: *jaws* there is no bony tip to the lower jaw; the *eyes* are very large (diameter greater than the snout length) one above the other or with the lower eye slightly in front of the upper; *snout* shorter; *scales* 87-93 along the lateral line; *fins* D 72-87; A 60-69.
The most easily recognisable characteristic of this fish is the uniform sandy colour of

the body and distinct dark spots towards the rear of both the dorsal and the anal fins.
Up to 41 cm.

A very deep water fish which is only rarely found above 100 m.

Wide-Eyed Flounder   p. 293 **pl. 204**
Bothas podas (Delaroche)
Platophrys podas Jordan & Goss

Mediterranean; east Atlantic north to Azores, south to Angola.

Body deep oval; *eyes* situated on the left side of the head. The lower eye is in front of the upper and they are separated by a very wide space. (In the female the lower eye is not completely in front of the upper and the space separating the eyes is not as great as in the male). The male fish has a spine directly in front of the lower eye and another on the snout; *mouth* small and situated on the left of the eyes; *scales* small 82-91 along the lateral line; *fins* the dorsal fin commences in front of the eyes. D 85-94; A 65-70.
The colour of the eyed-side is very variable and depends greatly on the colour of the bottom where the fish lives. They may be greyish, or brownish with lighter spots or uniform in colour. There are 1 or 2 distinct spots on the lateral line towards the tail. Blind-side bluish-white.
Up to 20 cm.

Usually found in shallow sandy areas where they feed on small animals. Breeding occurs during the summer.

Scaldfish   p. 293
Arnoglossus laterna (Walbaum)

East Atlantic north to Norway, south to Morocco; Mediterranean; North Sea; English Channel.
Body oval, slender; *eyes* situated on the left side of the head. The eyes are separated by a bony ridge and the lower eye is slightly in front of the upper; *mouth* situated on the

left of the eyes; *scales* very easily detached from the skin. 51-56 along the lateral line; *lateral line*, has a branch over the pectoral fin; *fins* dorsal fin starts in front of the eyes. The first few rays are partly free from the fin membrane. The pelvic fin on the blind side is very much smaller than that of the eyed side. Tail fin rounded. D 84-98; A 63-75.
Sandy brown or greyish with darker patches. The fins may have small dark spots. Up to 19 cm., usually around 10 cm.

The Scaldfish are most common in the Atlantic, Channel and North Sea where they are found on muddy or sandy bottoms from 10-200 m. In the Mediterranean they have been caught from 40-1000 m. but are most frequently found between 100-300 m. They feed on small bottom living animals. Spawning occurs in spring and summer.

Arnoglossus thori Kyle   p. 293
= *Pleuronectes grohmanni*

Mediterranean; east Atlantic from northern Ireland to Senegal; western Channel.

Body oval but fairly deep; *eyes* situated on the left of the head, close together, the lower eye slightly in front of the upper; *mouth* situated on the left of the eyes. Small it extends only to the first third of the lower eye; *scales* large 49-56 along the lateral line; *fins*, the dorsal fin starts in front of the eyes, the first few rays are longer and not attached to the fin membrane. The 2nd ray is elongate and has a membrane attached to it which gives the appearance of a fringe. The pelvic fin on the blind-side is much smaller than that on the eyed-side. D 81-91; A 62-67.
Eyed-side brownish or greyish with darker blotches and spots. Two of these blotches, one behind the curve of the lateral line and another towards the tail on the lateral line are more conspicuous than the others. There is also a narrow black stripe at the base of the tail fin and black spots scattered over the fins. There may be a dark spot on the blind-side of the pelvic fin.
Up to 18 cm.

Found on sandy, muddy and rough ground from 15 m. to below 100 m. Spawning occurs from March to July and possibly also during the winter in the Mediterranean. The adults feed on small bottom living animals, mainly invertebrates.

Arnoglossus imperialis (Rafinesque)   p. 293
= *Arnoglossus lophotes* Günther

East Atlantic north to Scotland; south to West Africa; western Mediterranean.

Body oval; *eyes* situated on the left side of the head; the lower eye slightly in front of the upper; *mouth* situated on the left of the eyes; *jaws*, lower jaw with a slight protuberance at the tip; *scales* large, detached very easily and often missing, 58-63 along the lateral line; *fins*, the dorsal fin commences in front of the eyes. In the male the 1st ray is fairly short but the next 5 or 6 rays are thickened and elongated and for the majority of their length free from the fin membrane. In females the first rays of the dorsal fin are slightly longer than the following rays and are free from the fin membrane at their tips only. D 95-106; A 74-82.
Eyed-side sandy or greyish with darker blotches. Male fish have a black spot on the pelvic fin.
Up to 25 cm.

This species is found on mud and sand in fairly deep water from 60-350 m.

Family:

PLEURONECTIDAE

This family includes most of the important flatfish including the **Plaice, Flounder, Lemon**

Sole and **Halibut**. The eyes are on the right side but very occasionally the eyes may be on the left side. This 'reversal' is relatively common in the Flounder (*Platichthys flesus*). The front gill cover has a free edge, the dorsal fin begins above the upper eye and the mouth is terminal. The egg yolk contains no oil droplet.

Dab   p. 298 **pl. 206**
Limanda limanda (L.)

Atlantic from the White Sea to Biscay, North Sea and western Baltic.
Body oval; *mouth* terminal, directed to the right of the eyes reaching back to the first 1/3 of the eye; *head* 1/4 body length; *lateral line* strongly arched over the pectoral fin; *scales* finely serrated along the margins giving a rough feel to the eyed-side, the blind-side has slightly rounded rear margins. D 65-81; A 50-64.
Yellowish-brown on the eyed side, often with indistinct blotches and small dark spots. Frequently found with scattered orange spots. Pectoral fin orange. The blind-side is white.
Up to 20 cm., rarely 40 cm.
One of the most common flat fishes of the north-east Atlantic. It prefers sandy shoals and banks from just beneath the surface to 150 m. or more. There appear to be no well defined spawning grounds but the young fish generally live on the bottoms only a metre or so beneath the surface. Spawning occurs from late winter to early summer, the earlier spawning being in the southern parts of its range. Whilst the Dab chiefly eats crustacea it will take almost any small animal it can find including small fish, echinoderms and molluscs. It feeds in a similar way to the Lemon Sole but strikes obliquely at its prey. This feeding method seems to be less suitable for catching polychaetes but is better for catching most other organisms.

Flounder   p. 298 **pl. 205**
Platichthys flesus (L.)

North Atlantic from the White Sea to Gibraltar, western Mediterranean. A related form *Platichthys flesus italicus* in the Adriatic.
Body oval; *jaws* terminal, directed to the right of the head, does not reach to beneath the eye; *lateral line* gently curved over the pectoral fin; *scales* small, without serrated margins. There is a row of hard warts running along the base of the dorsal and anal fins, similar warts either side of the lateral line; *fins*, tail fin square-cut. D 52-67; A 35-46.
Greenish or brownish, sometimes mottled with light or dark spots. Some fishes have large dull orange spots. Blind-side white.
Up to 30 cm., sometimes 50 cm.

The Flounder is particularly tolerant of differences in salinity and may be found high up river estuaries or even in completely fresh water. Alternatively they may be found in the sea down to 50 m. or more. It is most common in areas of rather low salinity such as the Baltic and is an important food fish there. In springtime they move offshore to their spawning grounds which are situated in the deeper part of their depth range. At first the eggs float near the surface, but as hatching and development proceeds they gradually sink deeper until by the time the typical adult flatfish form is reached the young fish mostly live on the bottom. The young fish which take up a bottom-living life soon after metamorphosis feed on a variety of small animals, especially crustaceans, the older fish feed on worms, shrimps and occasionally molluscs. Unlike the Plaice the teeth seem unsuitable for biting off food or crushing shells.

Plaice   p. 298 **pl. 135**
Pleuronectes platessa L.

North Atlantic from the White Sea to the Bay of Cadiz, western Mediterranean.

Body oval; *jaws* terminal, directed to the right and reaching back to beneath the first 1/3 of the eye; *lateral line* gently curved

Dab *Limanda limanda* p. 297

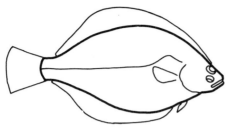

Curved lateral line;
back rough to touch.
20 cm.

Flounder *Platichthys flesus* p. 297

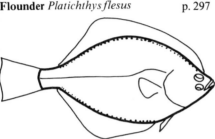

A row of bony warts at base of dorsal and
anal fin;
tail fin square-cut.
30 cm.

Plaice *Pleuronectes platessa* p. 297

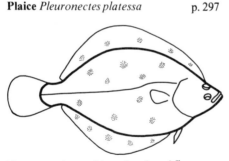

No warts at base of dorsal and anal fins;
bright orange spots.
60 cm.

Lemon Sole *Microstomus kitt* p. 299

Skin smooth and slimy;
warm mottled yellow-brown.
40 cm.

Witch *Glyptocephalus cynoglossus* p. 300

Long narrow body;
lateral line almost straight.
50 cm.

Long Rough Dab p. 300
Hippoglossoides platessoides

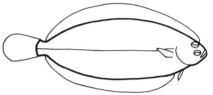

Large mouth directed to the right;
lateral line almost straight;
rough to touch.
50 cm.

Halibut *Hippoglossus hippoglossus* p. 300

Large mouth;
long, thick-fleshed body;
lateral line curved.
350 cm.

over the pectoral fin; *scales* have smooth margins; no bony warts at the base of the dorsal and anal fins. 4-7 warts between the beginning of the lateral line and the eyes. D 65-79; A 48-59.

Eyed-side brown with large conspicuously orange or orange-yellow spots. Blind-side white, sometimes with dark blotches.

Up to 60 cm., rarely up to 90 cm.

The Plaice is a very important commercial species, especially in the North Sea. It is also frequently seen by divers especially on sand and gravel bottoms. The orange spots on the back are very conspicuous underwater but when the nature of the bottom allows it the Plaice hides itself by flapping its fins in the bottom and thus covering itself with a deposit of bottom sediment. Where there is a tidal current the Plaice orientates itself pointing upstream and presses its dorsal and anal fins firmly onto the bottom.

In northern European waters the Plaice is found from the shore line down to about 100 m. or somewhat more; in the western Mediterranean it lives in much deeper water down to 400 m. Like other members of the family it is tolerant of brackish water and may move some way into estuaries to feed, but it is not nearly so tolerant of fresh water as the Flounder (*Platichthys flesus*).

North Sea Plaice generally spend the first year of their life in very shallow inshore grounds, but by their second year when they are about 15 cm. long they have begun to move into deeper water.

The females mature when they have grown to between 30 and 40 cm., the males when they reach 20-30 cm. The actual time they take to attain the lengths depends upon available food etc. but is usually between 4 and 5 years; sometimes as little as 2 years. Plaice spawn throughout their range usually on well defined spawning grounds. The biggest spawning grounds are in the deep water midway between the Thames Estuary and the Flemmish Bight and most Plaice inhabiting the southern North Sea appear to spawn there. Spawning is in winter or early spring. The eggs float in the upper layers of the sea, hatch after about 3 weeks

and the larvae develop into the typical flatfish form after about a further three weeks. By this time the young fish have sunk to the bottom and have normally drifted inshore to the shallow nursery grounds.

The young fish feed chiefly on small crustaceans, larvae etc. The older fish feed on shellfish such as cockles, razor shells and small scallops. The Plaice takes its food in a nearly horizontal position with its head raised slightly off the bottom. The cutting teeth in its jaws on the blind-side are well suited to bite off the protruding soft parts of molluscs whilst the crushing teeth in the pharyngeal region are very suitable for crushing up mollusc shells.

Lemon Sole; Lemon Dab   p. 298
Microstomus kitt (Walbaum)

North Atlantic from the White Sea to Biscay. North Sea and English Channel.

Body broad, oval; *head* small, only 1/5 body length; *jaws*, small directed to the right and not reaching the front of the eye; the lower jaw does not extend beyond the upper; *lateral line* gently curved over the pectoral fin; *scales* no bony warts, the scales have smooth margins, the skin is smooth to the touch; *fins*, tail rounded. D 85-97; A 69-76.

Eyed-side warm brown or yellow brown, mottled with darker brown, dull yellow or dull orange. Blind-side white.

Up to 40 cm., rarely 70 cm.

The Lemon Sole may be found on coarse sand or gravel bottoms from a few metres down to over 200 m. There are no well defined spawning areas. Spawning occurs from early spring to late summer; the earlier times being in the more southern waters. The food consists largely of bottom-living polychaete worms. Do not generally eat hard-shelled creatures but will bite off the protruding soft parts of shellfish. The feeding method of the Lemon Sole is rather specialised since its food is of the kind that is withdrawn back into a hole or into the protection of a hard shell at the least disturbance. When a suitable spot has been located the fish raises its head and the front part of its body off the bottom and remains in that position motionless in the water. Should a suitable worm or mollusc syphon present itself it strikes swiftly downwards on its prey. The rather small tube-like mouth and the cutting teeth in the jaws are apparently well suited for sucking worms from their burrows and for biting off the exposed ends.

Witch; Pole Dab   p. 298
Glyptocephalus cynoglossus (L.)
= *Pleuronectes cynoglossus*

Both sides of north Atlantic, in the east from the White Sea to Biscay, not in the North Sea or English Channel.

Body a long oval; *head* shallow pits on the underside; *mouth* small, directed to the right, reaching back to beneath the front edge of the eye; *eyes* large, longer than the snout length; *lateral line* directed slightly downward from the gill cover to the pectoral fin, from thence runs straight to the tail; *fins*, tail rounded. D 97-115; A 85-90.
Eyed-side greyish-brown, perhaps darker around the head, pectoral fin has a dark tip. Blind-side white.
Up to 50 cm.

The Witch lives on muddy bottoms from about 20 m. to 800 m. In Scottish waters, at least, the Witch begins to feed more actively in January and reaches one peak of feeding activity in March-April and another in the summer. In general it does not show pronounced breeding or feeding migrations although in Sweden it may come into shallow water to feed in the winter.
Like the Lemon Sole, it feeds chiefly on polychaete worms and its sucking and biting jaws seem particularly suitable for catching them. Unlike the Lemon Sole it probably feeds from a nearly horizontal position.

Long Rough Dab; Rough Dab   p. 298
Hippoglossoides platessoides (Fabricius)

North Atlantic from White Sea to Ireland, northern North Sea. A sub species *Hippoglossoides platessoides limandoides* (Bloch) in north-west Atlantic.

Body a long oval; *mouth* large directed to the right and reaching back to beneath the eye; *eye* as long or longer than snout length; *lateral line* almost straight but slightly curved over the pectoral fin; *scales* have serrated margins giving a rough feel; *fins* tail fin rounded. D 78-98; A 60-79.
Eyed-side greyish or reddish brown. Perhaps with dark mottling or spots. Blind-side white.
Up to 50 cm.

Lives on sandy or muddy bottoms and is tolerant of the low salinity in estuaries. In the northern part of its range it may come up the shore almost to the surface but is also found in depths approaching 400 m. Spawning is in spring and early summer, the later dates being in the north. It eats most kinds of small bottom living animals.

Halibut   p. 298
Hippoglossus hippoglossus (L.)

North Atlantic from the White Sea; Biscay, northern North Sea, rare in Baltic, absent in English Channel.

Body long and thick-fleshed; *jaws* large, directed to the right and reaching back to beneath the centre of the eye; *scales*, small, with smooth margins; *lateral line* strongly curved over the pectoral fin; *fins*, tail square-cut. D 98-106; A 73-80.
Eyed-side grey or olive-grey, blind-side white.
Up to 250 cm., sometimes up to 370 cm.

The Halibut is a very large, very palatable fish of great commercial importance. It is found on most kinds of bottom although perhaps not on mud. Since its depth range is from 100 m. to 1500 m. or more it is not likely to be seen by divers. It will only tolerate temperatures of 2.5°-8°C and moves into deeper water to escape rising temperatures. It is not an Arctic species since it cannot tolerate the sea temperatures of —1°C or lower that are found there. Spawns in the deeper part of its range in late winter or spring.

It is an active predator on fishes and will move into mid-water to feed. Its diet also includes bottom living crustaceans and molluscs.

Family:

SOLEIDAE

Eyes on the left side of the head. The egg has a single oil-globule in the yolk. The front gill cover has a free edge. Pelvic fin is not symmetrical. Mouth terminal.

Sole; Dover Sole
p. 302 **pl. 209**
Solea solea (L.)
= *Solea vulgaris* (Quensel)

East Atlantic north to Scotland; southern North Sea; English Channel; west Baltic; Mediterranean.

Body oval and elongate; *eyes* on the right side of the head; *snout* rounded; *mouth* small, semicircular in shape and situated on the lower edge of the body; *nostrils* small and tubular on the blind side; *scales* 116-165 along the lateral line; *fins*, 1 long dorsal fin which commences at a point midway between the upper eye and the tip of the snout; 1 anal fin similar to, but shorter than, the dorsal fin. Both the dorsal and anal fins are attached to the tail fin by a membrane. The pectoral fin on the blind-side is slightly smaller than that on the eyed-side. D 75-93; A 59-79.
Eyed-side brownish or greyish-brown with large, dark irregular blotches. The pectoral fin on the eyed side has a black spot at its tip which seems to intensify after death. The dorsal and anal fins have a white edge. Blind-side whitish. The colours of these fish vary greatly with the bottom and in some areas the fish markets are able to tell from which ground the fish were caught.
Up to 50 cm., usually only up to 30 cm. in the Mediterranean.

Found from shallow water down to 183 m.

either on, or half buried in, sand or mud. Frequently found in estuaries. During the winter there is a migration to deep offshore water, usually between 70-130 m. Feeding occurs mainly at night with peaks at dusk and dawn. The food consists of small bottom living animals. During the spring in the Atlantic the fish migrate to specific spawning areas which lie at a depth between 40-60 m. A number of these spawning grounds in the North Sea, Irish Sea and English Channel are known. Spawning occurs in winter in the Mediterranean. The young fish live in mid-water until they reach a length of between 15-18 mm. when they start their bottom living life. The Sole is of great commercial value and an excellent food fish.

Eyed Sole
p. 302
Solea ocellata (L.)

Mediterranean.

Body oval and elongate; *eyes* on the right side of the head; *snout* rounded; *scales* 70-75 along the lateral line; *fins*, 1 dorsal fin which commences at a point in front of the upper eye. The dorsal and anal fins are not attached to the tail fin. The pectoral fin on the blind-side is smaller and has fewer rays than that on the eyed-side. D 63-73; A 50-57.
Eyed-side variable in colour. Greyish-yellow, light-brown or chocolate brown with large dark spots on the front half of the

Sole *Solea solea*   p. 301

Black spot on pectoral fin.
51 cm.

Eyed Sole *Solea ocellata*   p. 301

4-5 dark, yellow centred and rimmed spots.
20 cm.

Thickback Sole *Solea variegata*   p. 303

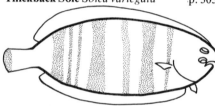

Dark cross bands on body;
upper eye diameter greater than that between the eye and the edge of the head.
22 cm.

Solenette *Solea lutea*   p. 303

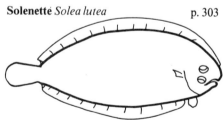

Every 5th or 6th ray of dorsal and anal fins dark;
upper eye diameter less than the space between the eye and the edge of the head.
No spot on pectoral fin.
13 cm.

Sand Sole *Pegusa lascaris*   p. 303

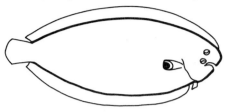

Black spot circled with white on the pectoral fin;
large rosette-like nostril on blind-side.
40 cm.

Solea kleinii   p. 304

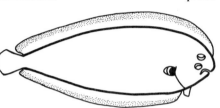

Black, orange and white pectoral fin;
anal and dorsal fins have dark edges.
32 cm.

Monochirus hispidus   p. 304

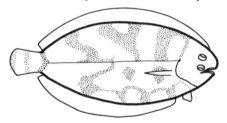

No pectoral fin on blind-side;
dark marbled patterning with a V-shape at base of tail fin.
14 cm.

body. The rear half has 4-5 dark spots each with a yellow centre and surrounded by a yellow ring. The base of the tail has a dark stripe and the tip of the tail fin has a fine white stripe. Blind-side white.
Up to 20 cm.

Usually found from 100-300 m. either on or half buried in mud or sand or amongst sea grass (*Posidonia*). Solitary fish are occasionally found in much shallower water.

Thickback Sole p. 302
Solea variegata (Donovan)
= *Microchirus variegatus* (Donovan)
Mediterranean; east Atlantic north to the English Channel and occasionally to Scotland.

Body oval and rather thick; *eyes* on the right side of the head. The upper eye is slightly in front of the lower eye and has a diameter greater than the space between the eye and the edge of the head; *snout* rounded; *scales* 70-92 along the lateral line; *nostrils*, small and tubular on the blind-side; *fins*, 1 dorsal fin which commences at a point level with the top edge of the upper eye. The anal and dorsal fins are well separated from the tail fin. The pectoral fin is small and much reduced on the blind-side. D 63-77; A 51-67.
The colour of the eyed-side is variable and ranges from greyish-yellow or reddish brown with about 5 darker cross bars which extend onto the dorsal and anal fins. There may also be other less clear, narrower stripes. The dorsal and anal fins may have dark rays scattered throughout their length. The rear edge of the tail fin has a dark stripe and the pectoral fin is dark. The blind-side is whitish.
Up to 22 cm., usually up to 14 cm. in the Mediterranean.

Found on fine or coarse sand. In the Atlantic they are most common between 35-90 m. but extend from 10-300 m. In the Mediterranean they range from 80-400 m., rarely shallower. Spawning occurs in the English Channel in spring and early summer at a depth between 55-75 m. and in the Mediterranean from February. The adults feed on small bottom living animals and are themselves an excellent foodfish.

Solenette p. 302 **pl. 207**
Solea lutea (Risso)
= Buglossidium luteum (Risso)

East Atlantic north to Scotland; English Channel; North Sea; Mediterranean.

Body oval and fairly elongate; *snout* rounded with the upper jaw slightly elongated and rounded to form a 'beak'; *eyes* on the right side of the head. They are small and the diameter of the upper eye is less than the space between it and the front of the head; *scales* 55-70 along the lateral line; *fins*, 1 dorsal fin which commences opposite the lower edge of the upper eye. There is a small membrane at the base of the dorsal and anal fins. The pectoral fins are small and much reduced on the eyed-side. D 65-78; A 50-63.
The colour of the eyed-side is variable but frequently yellowish or light brown either with or without darker blotches and spots. The dorsal and anal fins are sandy but every 5th or 6th (occasionally every 4th or 7th) ray is dark for the majority of its length.
Up to 13 cm., it is the smallest of the Soles.

This fish is common and may be found from 9-250 m. either on, or half buried in, sand. Spawning occurs in early summer in the English Channel and during spring in the Mediterranean. The young fish take up a bottom living existence at about 12 mm. in length. Adult fish feed on small bottom living animals. They have little commercial importance.

Sand Sole; French Sole p. 302
Pegusa lascaris (Risso)
= *Solea lascaris* Günther

East Atlantic north to Britain, occasionally along the west coast of Britain; Mediterranean.

Body oval and elongate; *snout* rounded with the upper jaw slightly longer than the lower; *eyes* on the right side of the head; *nostrils*, the front nostril on the blind-side is large (nearly the same diameter as the eye) and resembles a rosette; *scales* 96-140 along the lateral line; *lateral line* straight along the body but steeply curved above the eyes; *fins*, both the dorsal and anal fins have a membrane at their rear ends. The pectoral fin on the blind-side is only slightly smaller than that on the eyed-side. D 71-90; A 55-75.
The colour of the eyed-side is variable, greyish, yellowish or reddish-brown with small, irregular, darker spots scattered over the surface. The pectoral fin has a black spot circled with white at the extreme tip.
Up to 40 cm.

Found on or half buried in sand down to 48 m. It may be very common locally.

Solea kleinii (Bonaparte) p. 302

Mediterranean.

Body oval and elongate, gradually tapering towards the tail; *snout* rounded; *eyes* on the right side of the head, the upper eye slightly in front of the lower; *fins*, 1 dorsal fin which commences on the snout level with the upper edge of the upper eye. D 72-91; A 57-72.

Eyed-side very variable in colour, usually brownish with darker patches and small lighter spots. The edge of the dorsal and anal fins are dark. The pectoral fin has a black base, an orange centre and a white rear edge. Up to 32 cm.

Found on sandy or muddy bottoms from 20-100 m.

Monochirus hispidus Rafinesque. p. 302

Mediterranean; East Atlantic from Portugal to Morocco.

Body oval but tapers rather less than the other soles; *snout* rounded; *eyes* on the right side of the head; *scales* large and rather 'hairy'; *fins* the dorsal fin commences above the upper eye. There is no pectoral fin on the blind-side. D 50-58; A 40-45.

Eyed-side greyish or yellowish-brown with a dark marbled patterning which continues onto the dorsal and anal fins as a row of regular spots. There is a V-shaped stripe at the base of the tail fin.
Up to 14 cm.

This species may be found from 10-250 m. on sand, mud or sea grass (*Posidonia*) bottoms. Spawning occurs in late summer.

Order: DISCOCEPHALI = ECHENEIFORMES

The dorsal fin is transformed into a sucking disc on the top of the head.

Family:

ECHENEIDIDAE

The only family in the order Discocephali.

Common Remora; Shark Sucker **pl. 210**
Remora remora (L.)
= *Echeneis remora* L.

Uniform brown or greyish-brown;
17-19 pairs of ridges on the sucker;
tail fin slightly forked.
64 cm.

Mediterranean; East Atlantic to north Scotland; all warm oceans.

Body only slightly elongate and fairly stout; *head* large; *sucker*, there is a large sucker situated along the top of the head and the front part of the body. The sucker is about twice as long as wide and has between 17-19 pairs of ridges; *mouth* large and ends at the front edge of the eye; *jaws*, the lower jaw is longer than the upper and rounded; *fins*, the dorsal and the anal fins are opposite and similar to each other. Tail fin slightly curved. D 24-27; A 23-25.
Uniform dark brown or greyish-brown. Pectoral fins lighter.
Up to 64 cm.

Frequently found attached by the sucker on the top of the head to sharks and other large fish and mammals. They feed on the small crustacean parasites which infest the skin of the host, but they supplement their diet with other free living small crustacea and fishes. Spawning occurs in the early summer in mid Atlantic and in the late summer and autumn in the Mediterranean. Newly hatched fish live a free swimming life until they reach about 3 cm. in length when they attach themselves to a host fish, often another Remora.

Shark Sucker; Remora
Echeneis naucrates (L.)

Grey with a black longitudinal stripe flanked with white.
20-24 pairs of ridges along the sucker;
tail fin square cut.
65 cm.

Mediterranean; East Atlantic north to Portugal and all warm oceans.

Body longer and more slender than *Remora remora; snout* pointed; *sucker*, there is a large sucker situated along the top of the head and the front of the body. The sucker is three times as long as wide with 20-24 pairs of ridges; *jaws*, the lower jaw is much longer than the upper and has a fleshy, flexible tip; *fins*, the 2nd dorsal and anal fins are opposite and similar to each other, tail fin square-cut or slightly rounded. Pectoral fin pointed. D 32-40; A 31-39.
The back and underside are grey and the sides have a central, longitudinal black stripe flanked on each side by a white stripe.
Up to 65 cm., occasionally up to 1 m.

The habits of this fish are similar to those of *Remora remora.*

Order: PLECTOGNATHI = TETRAODONTIFORMES

Skeleton not completely bony with few vertebrae. The gill opening is a narrow slit situated in front of the pectoral fins. The bones of the upper jaw mostly fused sometimes forming a beak modified for grazing and scraping. The teeth may be distinct or absent.

Family:

BALISTIDAE

The teeth are implanted in sockets in the bone. The dorsal fin has three spines, the first is very strong and can be locked into position by a bony knob protruding forward from the base of the second spine.

Trigger Fish pl. 211
Balistes carolinensis (Gmelin)
= *Balistes capriscus* Valenciennes

Body diamond-shaped and laterally flattened;
outer rays of tail fin elongate.
25 cm.

Mediterranean; East Atlantic north to western Ireland; English Channel.

Body diamond shaped, deep but narrows just before the tail fin; flattened laterally, *snout* angular; *mouth* very small; *teeth* large and pointed; *lips* relatively large and fleshy; *gill slit* simple and situated immediately in front and above the pectoral fin; *scales* large and arranged in a mozaic pattern. There are about 2 large spines behind the gill slit; *fins*, 2 dorsal fins, the 1st fin consists of 3 spines. The 1st spine is large with a rough edge, the 2nd is smaller and close to the 1st. The 3rd spine is small and well separated from the 1st and 2nd spines. The 2nd dorsal fin and anal fin are similar in size and shape. Tail fin large with the outermost rays elongate. The pelvic fin consists of a large rough moveable spine. 1D III; 2D 27-28; A 25-27.

The colour is very variable and may be greenish, grey or brownish with green, blue or violet reflections on the back. The dorsal and anal fins sometimes have whitish or violet-black or bluish markings.
Up to 41 cm., usually around 25 cm.

Not strong swimmers, they swim by characteristic undulating movements of the dorsal and anal fins. Usually found in rocky or seaweed covered areas near coasts. The adults probably feed on small crustacea and molluscs which they find on rocks and on the sea bed. Spawning in the Mediterranean occurs during the summer when the sea reaches a temperature around 21°C. A hollow nest is excavated from the sand of the sea bed and in this the egg mass is laid. The adult fish guard the egg mass and ensure that it is aerated. The eggs hatch after about 2 days.

Family:

MOLIDAE

Body disc-shaped with a leathery skin and no scales. In the adults the tail fin resembles a wavy frill on the blunt rear part of the body.

Sunfish pl. 212
Mola mola (L.)

Throughout all the tropical and temperate oceans of the world.

Body almost rectangular and laterally flattened. There is no tail stalk; *snout* blunt; *mouth* small; *teeth* 1 beak-like tooth in each jaw; *gill slit* situated immediately in front of the pectoral fin; *skin* very thick and rough

Sunfish *Mola mola*

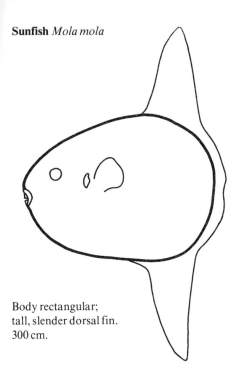

Body rectangular;
tall, slender dorsal fin.
300 cm.

but without scales; *fins* the dorsal and anal fins are short but tall. The tail fin is very narrow and attached to both the dorsal and anal fins, the rear edge undulates. D 16-20; A 14-18.

The back, fins and base of the tail fin are greyish, brownish or greenish; the flanks and undersurface are lighter.

Up to 300 cm.

Although these fish are most frequently seen floating near the surface in open water it is thought that many of these specimens are either old or dying; healthy fish probably live mainly in mid-water and possibly down to considerable depths. They feed on a large number of animals ranging from jellyfish, crustacea, echinoderms, fish and algae. Very little is known about their breeding habits but it is possible that the female may produce more than 300 million eggs.

Order: XENOPTERYGII = GOBIESOCIFORMES

Family:

GOBIESOCIDAE

Scaleless fishes with flattened bodies. There is a sucker modified from the pelvic fin. This is surrounded by a fold of skin supported in the front of the rays of the pectoral fin. One family only.

Blunt-nosed Clingfish; Blunt-nosed Sucker
Gouania wildenowi (Risso)   p. 308
= *Lepadogaster wildenowi* Risso

Mediterranean.

Body flattened sideways; *profile* curved; *head* rather flattened from above and wider than the body; *snout* blunt; *eyes* small separated by a space about 3 times the diameter of the eye; *gill slits* very prominent; *sucker* small and oval and situated under the front part of the body; *tentacle*, there is a slender branched tentacle in front of each eye; *fins*,

the dorsal and anal fins are continuous around the tail. The dorsal fin does not reach half way along the body. D 14-19; A 10-12.

Unlike all the other species of clingfish this one is uniform in colour, greyish-yellow, light brown or greenish-brown on the back and lighter on the belly.

From 4-7 cm.

Very little is known about this fish except that it is found attached to stones and rocks from very shallow water.

Blunt-nosed Clingfish *Gouania wildenowi*
p. 307

Uniform colour;
1 fin which continues around the tail;
branched tentacle in front of eye.
7 cm.

Small-headed Clingfish
Apletodon microcephalus

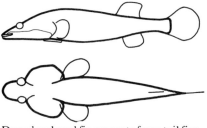

Dorsal and anal fin separate from tail fin;
greenish or brownish with white and dark
spots;
large canine type teeth at sides of jaws;
5 cm.
Female with light coloured throat.
Male with large cheeks.

Two-spotted Clingfish p. 309
Diplecogaster bimaculata

Dorsal and anal fins separate from tail fin;
anal fin smaller than dorsal;
reddish with yellow markings;
no canine-type teeth;
5 cm.
Male fish have a purple spot circled with
yellow behind the pectoral fin.

Shore Clingfish p. 309
Lepadogaster (lepadogaster) lepadogaster

Dorsal and anal fins attached to tail fin but
distinct;
large blue eye spots at the base of the head
on the back.
8 cm.

Connemara Clingfish p. 309
Lepadogaster candollei

Dorsal and anal fins not attached to the tail
fin; dorsal fin long;
male fish with red spots on cheeks and at
base of dorsal fin.
7.5 cm.

**Small-headed Clingfish; Small-headed
Sucker**
Apletodon microcephalus (Brook)
= *Lepadogaster microcephalus* Brook

Mediterranean; western English Channel
and northern Biscay; south-west coast of
Scotland.

Body flattened sideways; *head* flattened
from above; *sucker* situated under the front
part of the body; *snout* rounded; *cheeks*,
the male has enlarged rounded cheeks; *eyes*
fairly large, separated by a space about
twice the diameter of the eye; *nostrils*, the
front nostril has a small appendage on the
rear edge; *mouth* large; *teeth* there is a row
of 4-8 small incisor type teeth in the front of
the jaw and 1-3 larger canine type teeth at
each side. The teeth are characteristic of
this species; *fins*, the simple dorsal and anal
fins are short, opposite each other and nearly equal in size. They are separated from the

tail fin which is rounded. D 5-6; A 5-7.
Greenish, sometimes brownish, with lighter and darker irregularly arranged spots. The male usually has less of these spots and has been described as having a violet patch in the centre of the dorsal and anal fins and underneath the throat. The female has a light coloured throat.
Up to 5 cm.
Found down to 25 m. often amongst kelp (*Laminaria*) and Sea Grass (*Zostera*). Breeding occurs from May-September and the eggs are laid amongst the holdfasts of seaweeds.

Two-spotted Clingfish; Two-spotted Sucker   p. 308
Diplecogaster bimaculata (Bonnaterre)

East Atlantic north to Norway; western Channel, Mediterranean & Black Sea.

Body round but flattened sideways near the tail; *head* flattened from above; *profile* slopes from the tip of the snout to the back of the head; *snout* rather triangular; *sucker* situated under the front part of the body; *nostrils*; the front nostril has a small appendage on the rear edge; *mouth* rather large; *jaws* upper jaw slightly longer than the lower; *teeth* small and pointed and arranged in patches; *fins* the dorsal and anal fins are both separated from the tail fin. The anal fin is shorter than the dorsal. D 5-7; A 4-6.
This fish is extremely variable in colour, usually reddish on the back and sides and yellowish underneath with yellow, blue, and brown spots scatted over the surface. The yellow spots may be arranged in patterns. The male has a purple spot circled with yellow behind the pectoral fin.
Up to 5 cm.

This clingfish is found in deeper water than the other species; down to 55 m. in British waters and down to 100 m. in the Mediterranean. Usually they are found amongst sea grass or coarse bottoms often where there are plenty of shells. Breeding occurs during spring and summer. The eggs are laid in masses on the undersurface of shells or under stones and are guarded by the adults, particularly the male.

                                                         p. 308
Shore Clingfish; Cornish Sucker   pl. 132
Lepadogaster (lepadogaster) lepadogaster (Bonnaterre)
= *Lepadogaster gouanii* Risso

East Atlantic north to Scotland; western Channel; Mediterranean.

Body flattened sideways and back curved; *head* flattened from above and triangular in shape; *profile* slopes steeply from the snout to behind the back of the head; *sucker* situated underneath the front half of the body; *jaws* prominent and shaped rather like a duck bill. The upper jaw completely overlaps the lower; *lips* fleshy; *nostrils* with a large fleshy appendage on the rear edge of the front nostril; *fins*, the dorsal fin larger than the anal; both these fins are attached to the tail fin but there are distinct dips in the membrane. D 16-19; A 9-10.
Very variable in colour, yellowish, reddish or brownish with irregular brownish spots. There are two bright blue spots outlined in dark-brown, red or black on the back at the base of the head.
Up to 8 cm.

Found in very shallow water, from between the tides to about 1 m. They are frequently found in rock pools amongst stones and seaweeds. Breeding occurs during spring and summer. The eggs are golden in colour and are laid in masses which are fastened to the bottom of stones and guarded by the parent fish.

Connemara Clingfish; Connemara Sucker
Lepadogaster candollei Risso   p. 308
= *Mirbelia decandollei* Canestrini   **pl. 133**

East Atlantic north to Scotland, western Channel, Mediterranean; Black Sea.

Body flattened sideways in the rear half of the body; *profile* slopes gently from the tip of the snout to the back of the head; *sucker* situated under the front part of the body; *snout* rather long and pointed; *jaws* shaped rather like a duck-bill; *nostrils*, only a very small appendage present on the front nostril edge; *fins* dorsal fin long and not attached to the tail fin. D 13-16; A 9-11.

The colour of this fish is extremely variable. The ground colour may be reddish, brownish or greenish with red, brown or white dots, spots or stripes. Large specimens, probably the males, have 3 red spots at the base of the dorsal fin and other red spots on the head.

Up to 7.5 cm.

These fish are usually found on shores and around the low water mark amongst kelp, sea grass and stony bottoms. Eggs are laid under shells or stones where they are fixed by filaments on the eggs. They are laid down to about 30 m. in depth and guarded by the parent fish.

Order: PEDICULATI = LOPHIIFORMES

The dorsal fin composed of a few flexible rays of which the first is placed on the head and generally ends in some sort of tassel or bulb; in some deep-sea species this bulb is luminous. In some deep water forms the male is dwarfed and lives a parasitic existence on the female.

Sub-order: LOPHIOIDEA

Pelvic fins present (toothed), males not dwarfed, frontal bones in contact for most of their length.

Family:

LOPHIIDAE

The only family in the sub-order.

Angler Fish; Frogfish; Monkfish **pl. 134**
Lophias piscatorius (L.) **pl. 199**

Throughout Atlantic, Mediterranean; English Channel; North Sea.

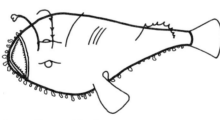

Large flattened body;
very large semicircular mouth;
large spine with a fleshy lobe at the tip just behind the upper lip.
198 cm.

Body, front part of body flattened from above but gradually tapers to a normal fish-shaped rear half; *head* large, wide and flattened from above; *mouth* very large, semicircular in shape; *jaws*, lower jaw longer than upper; *teeth* curved, irregularly sized and spaced; *eyes* small and situated on top of the head; *gill slits*, 2 large openings immediately behind the pectoral fins; *skin*, there are no scales but a number of fringed lobes which run around the sides of the head and the body. The skin is very loose; *fins*, there are a number of isolated rays which run down the mid-line. The 1st ray, which is the longest is situated just behind the lip of the upper jaw and ends in a branched fleshy lobe. The 2nd ray is shorter but fringed and situated in front of the eye. The 3rd ray is found someway behind the other two. There are 2 dorsal fins, the 1st with 3 rays joined by a very short membrane. The 2nd is longer and opposite and similar to the anal fin. The pectoral fins are large; the pelvic fins are small and situated on the undersurface of the head. D I + I + I + III; 2D 10-13; A 9-11.

Brownish, reddish or greenish-brown with darker blotches. Undersurface white, except for the rear edge of the pectoral fin which is black.

Up to 1.98 m.

Common, range from 18-550 m. They are normally encountered either on or half buried in sandy and muddy bottoms, but may also be found amongst seaweeds and sandy areas between rocks. They feed on smaller fish (flatfish, haddock, dogfish etc.) which they attract by twitching the fleshy lobe at the end of the 1st dorsal ray. They do, however, feed on practically any other bottom living animals, and may even make short trips off the bottom and have been recorded as attacking sea birds on the surface. Day (1888) states that they are able to creep along the bottom using their pectoral fins and are therefore able to move without agitating the water.

Spawning occurs during late winter, spring and summer in very deep water, possible as deep as 180 m. The eggs are laid in large ribbon like sheets up to 9 m. long and 90 cm. wide. After a free living existence the young take up a bottom living life at about 5-6 cm. The Angler Fish is of no commercial value.

Index

318